Sport, Development and Environmental Sustainability

This is the first book to consider the intersections of sport, international development and environmental sustainability. It explores the tensions between sport's potential contribution to the environment and its rather poor record to date.

Bringing together a diverse group of scholars who approach the topic from various disciplinary and theoretical perspectives, the book provides both critical and optimistic perspectives on the place of sport in sustainable development. Chapters examine and question how and whether sport contributes to sustainable development on an international scale. Attention is also paid to the place and role of Indigenous knowledge in sustainable Sport for Development, particularly as an alternative to modernization and/or in support of reconciliation with Indigenous peoples.

Sport, Development and Environmental Sustainability is important reading for academic researchers, students and policy-makers in the fields of kinesiology, sport studies, sport sociology, leisure studies, sport management, sport media, physical cultural studies, environmental studies and sustainability and international development studies.

Rob Millington is Assistant Professor in the Department of Kinesiology at Brock University, Canada. His research focuses on how international NGOs such as the United Nations and International Olympic Committee mobilize sport for development in policy and practice, in both historical and contemporary contexts. His postdoctoral work considered what role, if any, sport can play in meeting the UN's Sustainable Development Goals.

Simon C. Darnell is Assistant Professor in the Faculty of Kinesiology and Physical Education at the University of Toronto, Canada. His research focuses on the relationship between sport, international development and peace-building, the development implications of sports mega-events and the place of social activism in the culture of sport. He is currently an Associate Editor of the *Sociology of Sport Journal* and sits on the editorial boards of five other journals, including the *Journal of Sport for Development*. He has served as a guest editor for issues of *Third World Quarterly* and *Qualitative Research in Sport, Exercise and Health*.

Routledge Studies in Sport Development
Series Editors

Richard Giulianotti
Loughborough University, UK and University of Southeast Norway

B. Christine Green
George Mason University, USA

The *Routledge Studies in Sport Development* series showcases high-calibre work within the vibrant, diverse and rapidly expanding field of sport and development. It includes books in two broad areas: firstly, the *development of sport*, focusing on the various ways in which sport is delivered, for example through building sport facilities, training coaches and athletes, improving sport performance, increasing public participation in sport and strengthening the governance, management, marketing and delivery of sport, and secondly, *Sport for Development and Peace*, examining how sport is used for different non-sporting social benefits, such as peace-building and conflict reduction, health education, gender empowerment, community development, tackling crime, improving education, promoting 'positive youth development' and advancing the social inclusion of marginal populations. The series is committed to diversity in theory and method, is multi-disciplinary in approach and includes work centring on local, national and transnational issues and processes, and on the global North and/or South.

Available in this series:

Routledge Handbook of Sport for Development and Peace
Edited by Holly Collison, Simon C. Darnell, Richard Giulianotti and P. David Howe

Sport, Development and Environmental Sustainability

Edited by Rob Millington and Simon C. Darnell

Routledge
Taylor & Francis Group

LONDON AND NEW YORK

First published 2020
by Routledge
2 Park Square, Milton Park, Abingdon, Oxon OX14 4RN

and by Routledge
52 Vanderbilt Avenue, New York, NY 10017

Routledge is an imprint of the Taylor & Francis Group, an informa business

© 2020 selection and editorial matter, Rob Millington and Simon C. Darnell; individual chapters, the contributors

British Library Cataloguing-in-Publication Data
A catalogue record for this book is available from the British Library

Library of Congress Cataloging-in-Publication Data
A catalog record has been requested for this book

ISBN: 978-0-815-35613-4 (hbk)
ISBN: 978-1-351-12862-9 (ebk)

Typeset in Goudy
by Swales & Willis, Exeter, Devon, UK

Contents

Contributors

Kyle S. Bunds is an Assistant Professor in the Department of Parks, Recreation, and Tourism Management at North Carolina State University, USA. His research and teaching examine the connection between sport and the environment generally, and sport, water and air pollution more specifically. His work, which is primarily grounded in political economic theory, has been published in numerous academic journals, including *Sport Management Review; European Sport Management Quarterly; Sport in Society; Critical Studies in Media Communication; Communication, Culture, & Critique; Cultural Studies Critical Methodologies;* and *Water Resources: IMPACT.* In addition to his scholarship, Kyle has also served as guest editor for a special issue on political economics for the *Journal of Amateur Sport*, and a special issue on sport, physical culture and the environment in the *Sociology of Sport Journal.*

Tanya Halsall is a post-doctoral research fellow at the Royal's Institute of Mental Health Research affiliated with the University of Ottawa, Canada. Her primary research areas are in positive youth development, program evaluation and community-based research with First Nations, Métis and Inuit (FNMI) youth. Her specific research interests are in community-based participatory research and evaluation of sport-based programming for youth that promotes engagement, leadership and well-being. She has also been involved in evaluating system-level initiatives in child and youth mental health at the regional, provincial, national and international levels. These collaborative initiatives have applied youth engagement strategies and placed a focus on the promotion of wellbeing in FNMI youth.

Michael Heine is an Associate Professor in the School of Kinesiology at Western University, Canada, and the Director of the International Centre for Olympic Studies, specializing in social theory, qualitative methods and historical sport sociology. His specific research interests include Indigenous Games in sub-Arctic and Arctic Canada, discursive analyses of economic processes in sport, and the Olympic commodification of civic spaces.

Daniel Henhawk is an Assistant Professor in the Faculty of Kinesiology and Recreation Management at the University of Manitoba, Canada. His research revolves around sport, recreation and leisure in the context of Indigenous communities. More specifically, he is interested in the issues that surround colonization, decolonization, Indigenization, self-determination and sovereignty. His research interests also broadly include narrative inquiry and the use of stories and storytelling as a means to examine the tensions between historical and contemporary understandings of leisure, sport and recreation in relation to Indigenous ways of knowing and being.

Tanya Forneris is a Senior Instructor in the School of Health and Exercise Sciences at the University of British Columbia Okanagan, Canada. Her research interests include community-based program development and evaluation, positive youth development through sport, and post-secondary student wellbeing.

John Karamichas is Lecturer in sociology and green criminology in the School of Social Sciences, Education and Social Work and Fellow of the Centre for the Study of Risk and Inequality at Queen's University Belfast. His publications include *The Olympic Games and the Environment* (2013) and *Olympic Games, Mega-events and Civil Societies: Globalization Environment, Resistance* (with Graeme Hayes, 2012).

Kyoung-yim Kim is an Assistant Professor of the Practice of Social Science at Boston College, USA and teaches in the Departments of Sociology and Women's and Gender Studies. Her research focuses on inequality, labor migration, decolonizing methodologies, environmental sustainability and policy, with particular attention paid to postcolonial and transnational feminist approaches. Her work has dealt with three major geographic areas: the United States, South Korea, and Japan. Her next project involves biocolonialism in the transnational golf industry and professionalization of women's sport.

Brad Millington is a Senior Lecturer in the Department for Health at the University of Bath, UK. His research is focused mainly on two areas: sport media and technology, and sport's relationship with the environment. Brad has authored two books: *Fitness, Technology & Society: Amusing Ourselves to Life* (2018) and *The Greening of Golf: Sport, Globalization and the Environment* (2016, with Brian Wilson).

Richard Norman is a PhD candidate in the Department of Recreation and Leisure Studies at the University of Waterloo, Canada. His research focuses on the experiences of participants related to issues of inclusivity, hegemony and diversity in sport. Richard's dissertation explores the role of narrative in probing notions of dominance in the sport of curling, and the implications for participant growth related to new Canadians and people of colour.

Victoria Paraschak is Professor in the Department of Kinesiology at the University of Windsor, Canada. Her research focuses on Indigenous peoples and sport, most recently within a strengths-and-hope perspective. She has facilitated several workshops using a strengths perspective to help improve conditions for Indigenous sport in Canada. Vicky previously worked with the Government of the Northwest Territory's Sport and Recreation Division as a policy officer, and as a consultant facilitating various strategic Directions Conferences. Her research has focused on working with others to address the Truth and Reconciliation Commission's call to action #87 since 2015; projects have included consolidating and expanding elite Indigenous athlete entries on Wikipedia, and creating indigenoussporthistory.ca.

Carolyn Prouse is an Assistant Professor in the Department of Geography and Planning at Queen's University, Canada. Her PhD dissertation work focused on technologies of 'slum-upgrading' in Rio de Janeiro and she continues to explore the urban political ecologies of both productive and reproductive labours. In general, Carolyn's research interests fall at the intersection of post/decolonial urbanism, feminist studies, and critical race theories.

Mitu Sengupta is Professor in the Department of Politics and Public Administration at Ryerson University, Canada. She has a PhD in Political Science from the University of Toronto, and a Master of Arts and Bachelor of Arts (Honours) in Political Science from McGill University. She has published widely on Indian market liberalization and development, on labour and migration in India, and on the politics of sporting events and film. Mitu's more recent work focuses on normative concerns in international development and international relations, such as the quest to create universal development goals (like the UN Sustainable Development Goals) and fair climate change and trade agreements. She is interested, in particular, in India's interventions in these global arenas, and their relationship with domestic normative debates and social movement discourses.

Larry Swatuk is Professor in the School of Environment, Enterprise and Development at the University of Waterloo, Canada. His current research interests focus on the unintended negative consequences of climate change adaptation and mitigation interventions, a concept he labels 'the boomerang effect'. He is author of *Water in Southern Africa*. Dr. Swatuk is also Adjunct Professor of International Development, St. Mary's University, Halifax; External Research Fellow, Centre for Foreign Policy Studies, Dalhousie University, Halifax; Senior Research Fellow, Bonn International Centre for Conversion, Bonn, Germany; Extraordinary Professor, Institute for Water Studies, University of the Western Cape; and a Research Fellow of both the Balsillie School of International Affairs and the Water Institute in Waterloo, Ontario, Canada.

Devra Waldman is a PhD candidate in the School of Kinesiology at the University of British Columbia, Canada, and a Liu Scholar at the UBC Liu Institute for Global Issues. Her research focuses on international development, urban studies and sport.

Gavin Weedon is Senior Lecturer in the Sociology of Sport, Health, and the Body in the School of Science and Technology at Nottingham Trent University, UK. His research explores embodied practices from a range of fields, theories and disciplines, with a focus on nature-society relations, political economy and consumerism, and the politics of embodiment. His current projects examine the political ecology of protein supplementation (a collaboration with Dr. Samantha King, Queen's University, Canada), telomere biology and the Anthropocene, and the 'back to nature' fitness movement.

Brian Wilson is a sociologist and Professor in the School of Kinesiology at the University of British Columbia, Canada, and Director of UBC's Centre for Sport and Sustainability. He is co-author of *The Greening of Golf: Sport, Globalization and the Environment* (2016, with Brad Millington) and author of *Sport & Peace: A Sociological Perspective* (2012) and *Fight, Flight or Chill: Subcultures, Youth and Rave into the Twenty-First Century* (2006). He is co-editor of *Sport and Physical Culture in Canadian Society* (2020, with Jay Scherer) and *Sport and the Environment: Politics and Preferred Futures* (forthcoming, with Brad Millington). His research and writing focus on sport, social inequality, environmental issues, media, peace, social movements, and youth culture.

Chapter 1

Introduction

Sport, development and environmental sustainability – issues and controversies

Rob Millington and Simon C. Darnell

Introduction

In October 2015, the United Nations (UN) announced the 17 Sustainable Development Goals (SDGs), designed to form the focus of global development through 2030. The SDGs marked a shift in development policy in at least two ways: first, the goals foregrounded environmental sustainability as a primary issue of international development, and made clear the need for environmental protection and remediation. Included in the SDGs are, for example: access to sustainable energy (SDG 7), sustainable industrialization (SDG 9), resilient and sustainable cities (SDG 11), sustainable consumption (SDG 12), action to combat climate change (SDG 13), conservation of the world's oceans (SDG 14) and protection of the forests and biosphere (SDG 15). Second, the SDGs directly codified sport within international development policy for the first time, marking something of a point of culmination for the Sport for Development and Peace (SDP) sector, the loose amalgam of organizations and stakeholders that advocate for sport in the service of international development and peacebuilding (Giulianotti, 2012). Indeed, beginning in the early 2000s, sport grew significantly within the development sector, thanks in large part to the UN's institutionalization of sport for development and its creation of the position of Special Adviser to the UN Secretary-General on Sport for Development, and the United Nations Office on Sport for Development and Peace (UNOSDP).[1]

During this time, organizations like the UN, as well as other SDP proponents, increasingly connected the use of sport to a range of development objectives, including poverty alleviation, HIV/AIDS education and gender equity. Notably, the environment also appeared to assume an increasingly important role in SDP, with the UN positing that sport could raise "awareness towards climate protection and … stimulate enhanced community response for local environmental preservation" and might even "make significant contributions to combat climate change" (United Nations, 2016, p. 14).

Accompanying this growth of the SDP sector was a series of rejoinders, particularly from the field of sport sociology, in which scholars critiqued the politics of evidence in SDP (Coalter, 2007; Nicholls, Giles, & Sethna, 2011), the influence of corporatization on the sector (Hayhurst, 2011; Hayhurst & Szto, 2016; Levermore, 2011), and even SDP's complicity with neo-colonialism and social reproduction (Darnell, 2007; Hartmann & Kwauk, 2011). To date, however, there has been comparatively little critical analysis of environmental issues within the burgeoning SDP sector, or the role of sport with regard to environmentally sustainable development on an international scale. This oversight is significant given that the environment has been tied to matters of international development for decades, and has long been a focus of scholars within development studies (cf. Kirsch, 2010). Furthermore, as climate change has become a matter of global concern, environmental management and politics have emerged as a central focus of development policy, particularly given that "developing" polities are particularly vulnerable to the effects of climate change (cf. Adger et al., 2003).

This research gap in the relationship between sport, environmental sustainability, and international development was the impetus for hosting a research symposium in June 2017 at the University of Toronto entitled: *Sport and Sustainable Development: Setting a Research Agenda*. The main goal of the symposium was to bring together scholars in the fields of critical sport studies, development studies, political science and environmental studies to discuss the role that sport can (and should) play in efforts towards sustainable development. The subsequent aims of the symposium were threefold:

- To assess the current state of the field, and explore how environmental sustainability has been, or could be, included within the burgeoning field of SDP research;
- To consider the theoretical, political and historical points of intersections between these different disciplines;
- To identify areas of future research, and begin to develop a research paradigm for the field.

The collection of essays that comprise this book emerged from the symposium. They represent a diversity of scholarly assessments of the sport-development-environment nexus, and attempt to provide both a series of critical insights into the topic, while also proposing questions and concerns still to be explored. Through this Introduction, we aim to contextualize the chapters that follow by providing a discussion of the key themes that emerged from the symposium, as well as an overview of the tensions and controversies surrounding sport's potential within efforts of sustainable development. We offer an assessment of the "state of the field" before discussing three central themes that we see as requiring further and ongoing attention within the SDP literature: the role that sport can or should play in promoting sustainable development, with a particular emphasis on environmental issues; the role that non-governmental organizations and federations from the field of sport might take up in this regard; and issues related to the intersection of sport and eco-justice. We conclude the Introduction by identifying some future areas of research as well as previewing the chapters that comprise the book.

Sport-development-environment: the state of the field

As noted, research into the relationship between sport, international development and the environment is both timely and called for because of the growth of the SDP sector in recent years. Beginning in the early 2000s, intergovernmental organizations like the UN began to recognize sport's potential in contributing to international development. Over a relatively short period, hundreds of organizations began to mobilize sport in the service of international development, including non-governmental organizations like the Canadian-based Right to Play, corporations such as Nike and sporting federations like the International Olympic Committee, which in recent years have sought to connect the Olympic Games to broad-based development strategies for host nations.

However, despite this growth of SDP, it is only recently that environmental protection and remediation have been emphasized within the sector, and it is still somewhat rare for SDP organizations to focus primarily on the environment. As mentioned, this is curious from the perspective of international development studies, where the environment has been of crucial importance since at least the 1970s and in which global climate change is seen as nothing less than a threat to sustainable human life. The recent attention paid to the environment in SDP is also conspicuous given that the United Nations Environmental Programme used the popularity of sport in the 1990s to "promote environmental awareness and respect for the environment among the public, especially young people" and established environmental guidelines for sporting events such as the Olympic Games (Jarvie, 2012, p. 266). These guidelines eventually informed the International Olympic Committee's environmental mandate, as well as programs and activities designed to "contribute to raising awareness about the importance of sustainable development in sport" (IOC, 2014), particularly for global South nations.

In turn, the recent attention paid to sustainability through frameworks like the SDGs can perpetuate the idea that the environment is a relatively new feature of international politics, when in fact the SDGs are but the *latest* attempt to bring attention to sustainable development. The 1972 *UN Conference on the Human Environment*, for instance, sought to bring together developed and developing countries to establish a global consensus on environmental issues. The conference resulted in the Stockholm

Declaration, a set of 26 principles and 109 recommendations concerning the environment and development, including the need to safeguard natural resources and produce renewable forms of energy, and a mandate to contain and prevent pollution, while stressing the importance of environmental education (United Nations, 1972). In addition, one of the key themes to emerge from the conference was that development was central to environmental remediation, and that developing countries required assistance in this regard, particularly with environmental protection. Fifteen years later, the UN convened the Brundtland Commission to rally countries in collective pursuit of environmentalism in development. The Brundtland Commission popularized the term "sustainable development," defining it as "development that meets the needs of the present without compromising the ability of future generations to meet their own needs" (WCED, 1987, p. 41). This approach employed two key concepts:

> The concept of "needs", in particular the essential needs of the world's poor, to which overriding priority should be given; and the idea of limitations imposed by the state of technology and social organization on the environment's ability to meet present and future needs.
>
> (WCED, 1987, p. 41)

Whereas previously the environment was understood to be separate from human action and development was typically defined within political or economic goals, the Brundtland Commission was instrumental in linking the two concepts. Five years later, the UN hosted the *Rio de Janeiro Earth Summit* to focus on climate change, and largely set the stage for the Kyoto Protocol and eventually the 2016 Paris Climate Accord.

While the concern for the environment within approaches to international development is not new, the threat posed by climate change has significantly heightened the focus afforded to sustainability within international development policy and research. This has arguably necessitated a shift in development thinking and efforts, away from so-called "big picture" development policies, like efforts to "make poverty history" (see Cooper, 2007; Grant, 2015) or consumer-focused attempts to "make development sexy" (Cameron & Haanstra, 2008; Richey & Ponte, 2008), and towards more focused and coherent attention paid to sustainability. While a general withdrawal of the state, decline in multilateralism and concomitant rise in bilateralism has resulted in the NGO-ization of development, in which the number of actors with fragmented goals and competing policies has proliferated, the SDGs nonetheless represent an attempt to foreground the importance of the environment within the field of international development. Amidst this, the specific inclusion and recognition of sport in the SDGs was of clear significance. It was against this backdrop that participants in the *Sport and Sustainable Development: Setting a Research Agenda* symposium took up the question of sport, development and environmental sustainability. In what follows, we outline the three key themes which emerged from this two-day meeting.

What role can/should sport play in sustainable development?

The first and most pressing theme was the need to consider what role sport can or should play in efforts of environmentally sustainable development. Despite the efforts of organizations like the UN to connect sport to environmental protection and remediation, there remains a significant gap in the academic literature regarding the efficacy of these claims. Much of the support for sport's potential to contribute to sustainable development policies and practices is based on *a priori* and often vague assumptions about the "power" of sport as a transformative force (see Donnelly, 2011; Hayhurst, 2009). Conversely, the existing research on sport and the environment has been almost entirely critical of sport's impact on the natural environment, particularly the effects of hosting sporting events on natural landscapes, the levels of waste production and carbon footprints that sport produces, and the displacement of local habitats (and people) in order to build sports facilities (Bale, 1994; Chernushenko, 1994; Wilson & Millington, 2013).

Particular sports have received a great deal of attention regarding their impact on ecosystems. For example, the deleterious environmental effects of golf course construction and management have long been a focus for academics and activists who have drawn attention to issues of deforestation, clearing of vegetation, application of pesticides and over-use of water supplies (Briassoulis, 2010; Millington & Wilson, 2014, 2015; Neo, 2010). In response to the tide of environmental opposition since the 1960s, the golf industry took up an "environmentally conscious" approach in order to promote golf as a green and environmentally friendly sport that allows people to connect with nature (Briassoulis, 2010; Millington & Wilson, 2015). As we have noted elsewhere (Millington, Darnell, & Millington, 2018), this "environmentally friendly" shift in golf mobilized scientific discourses and presented a way forward for the golf industry whereby "technological advancement would not only afford cleaner and safer (or, at least, less risky) approaches to golf course development and maintenance, but also demonstrate the ability of non-state actors to self-monitor, thus erasing the need for oversight or burdensome legislation" (p. 11). This approach has most often been referred to as "ecological modernization," recognized for its claims that the growth of the sport sector is compatible with environmental sustainability. However, many of the environmental efforts within the golf industry, similar to those of other corporate entities, have also been accused of engaging in "greenwashing," whereby sport organizations and corporations attempt to *market* themselves as environmentally conscious and their products as sustainable without changing, or indeed ending, the policies and practices that fundamentally harm the environment (Lenskyj, 1998; Miller, 2016, 2018). In many ways, the golf industry's promulgation of ecological modernization (and greenwashing) is emblematic of the sport industry, *writ large*.

With this history and track record in mind, there are questions as to what role, if any, sport can reasonably play in promoting environmentalism or achieving sustainability. Some stakeholders continue to argue that the global sport industry can and/or might raise awareness of environmental issues while mobilizing its immense resources in the service of environmental protection and remediation. Such claims should not be dismissed out of hand. Given its reach and global visibility, it may indeed be reasonable to pursue opportunities for sport to act as an "attractor discourse" for sustainable development, one that aligns with governmental policies and stakeholder needs (Mol, 2010). Further, on the production side, multinational sports equipment corporations could conceivably play a role in implementing new environmental standards and initiatives, particularly in the global South. Sport might also offer a means of attracting youth to learn about the environment through outdoor education, through the building of partnerships with environmental organizations, and/or by implementing environmentally conscious travel schedules (see Halsall and Forneris, Chapter 10 in this collection).

Such possibilities extend to urban settings as well. The construction of new stadiums, particularly the building of multi-sport facilities that take into consideration proximity to environmentally sensitive areas and access to public transportation, could be achieved in sustainable ways that make use of environmental impact studies and utilize renewable energy sources. Public-private partnerships might also offer an opportunity to ensure that a variety of stakeholders' needs and perspectives are met with regard to sport and sustainable development (see Trendafilova, Babiak, & Heinze, 2013).

However, despite such opportunities, the specter of greenwashing and the limits of ecological modernization continue to loom over the global sports industry. The question remains whether sport advocates and stakeholders are willing and/or able to make the kinds of changes and commitments necessary to achieve actual environmental sustainability. In addition is the question of which organizations can or should be responsible for such efforts, which is discussed in the next section.

What is the role for sport organizations in sustainable development?

The second theme to emerge from the symposium concerns the role that sports organizations, and particularly international non-governmental organizations (INGOs), can or should play in environmental sustainability, especially on an international scale. For example, given the environmental demands of sports mega-events, and the putative "need" to build new sporting facilities for the Olympic Games and FIFA World Cup every two to four years, sports organizations have a significant environmental responsibility. While sport in general, and sports mega-events in particular, have been posited as able to provide a platform for discussing environmental issues (see Mol, 2010), it is still more likely that sports events will exacerbate, rather than mitigate, environmental degradation.

The International Olympic Committee (IOC), in particular, has been subject to criticism regarding the sustainability of the Olympic Games, with scholars and activists pointing to the environmental damages that often result from hosting this event. This continues despite the IOC's efforts to frame the event as environmentally friendly. Critiques have focused on increases in waste production, pollution, energy consumption and the reconfiguration of landscapes (e.g. the construction of stadiums) that the Games routinely require (Hayes & Karamichas, 2012; Karamichas, 2013). In this regard, John Karamichas (2013) has argued that there is little evidence to suggest that hosting sports mega-events can contribute to environmental sustainability in any meaningful way, and that the games in fact compound environmental degradation, even though notions of environmental "remediation" or "development" have specifically informed bids to host sport mega-events in Mexico (Bolsmann & Brewster, 2009; Brewster, 2010), South Africa (Cornelissen, 2011), China (Qing, 2010; Zhang & Silk, 2006) and Brazil (Christopher Gaffney, 2010; Darnell, 2012; Millington & Darnell, 2014; Millington, Darnell, & Millington, 2018). This is all exacerbated by the fact that host cities in developing countries are often in a weaker position to deal with the associated costs of staging the Games, including damage to the environment (Black & Van Der Westhuizen, 2004; Cornelissen, 2010).

Similar to the golf industry's efforts to frame itself as "environmentally friendly," the IOC has been forced to respond to criticisms of its environmental record. Facing critiques regarding the detrimental effects of hosting the Olympic Games since at least the 1970s, the early 1990s were a turning point in the IOC's position on environmental issues. In 1992, the Albertville Olympic Games were environmentally disastrous, particularly the destruction of the natural landscape to create ski slopes, and drew international condemnation (Cantelon & Letters, 2000). At the time, the IOC was eager to align itself with Norway's bid for the 1994 Games, led by Prime Minister Gro Harlem Brundtland (the head of the Brundtland Commission, discussed above), and as a result, the Norwegian bid served to articulate an environmental vision for the Games that ultimately served the IOC's interests. Eventually, the IOC introduced an environmental requirement for bid cities, created a Commission on Sport and the Environment, and positioned itself as a presumptive leader for the environmental movement. Since then, the IOC has continued to promote its leadership in the environmental sector, with a particular focus on the sustainability legacy the Olympics afford host cities and countries.

It is worth noting that FIFA, the governing body of global football/ soccer, has been subject to similar critiques over its environmental record, with activists and scholars calling attention to the limited sustainability of the World Cup, focusing in particular on the massive infrastructural developments the event often requires. The 2002 World Cup in Korea and Japan, for example, necessitated the construction or remodeling of as many as 20 stadiums, many of which have since fallen into disuse (Gaffney, 2013). In a response that mirrors that of the IOC, FIFA dedicated corporate social responsibility directives to making the World Cup more sustainable and "green." However, as Christopher Gaffney (2013, p. 3928) argues, FIFA's initiatives are less ambitious than those of the IOC, with the stated goal of "minimizing the negative impacts of the World Cup," a tacit admission that regardless of their efforts, the event has significant negative impacts on the environment.

Overall, then, sports organizations and events, often led by non-elected INGOs like the IOC and FIFA, continue to occupy an important but ambivalent position with regard to sustainable development on a global scale. If these powerful organizations are genuinely concerned about the environmental impact of the mega-events which they oversee, then serious consideration needs to be given to the ecological footprint, and the pollution and changes to environmental landscapes, that such events produce and/or require. Again, these issues are of particular concern for developing nations in the global South given the recent tendency to connect hosting of sport mega-events in such countries to broad-based development strategies. The 2008 Beijing Olympics, 2010 South Africa World Cup, 2012 Sochi Olympics, 2014 Rio World Cup, 2016 Rio Olympics and the 2022 World Cup in Qatar all serve as recent examples.

In sum, sports organizations and event organizers should be held accountable for their environmental practices and footprints. Further, the current model of hosting sports mega-events, in which cityscapes are reinvented and thousands of people flown to a single location requires reconsideration, now more than ever. Absent of this, organizations like the IOC and FIFA can continue to mobilize sport for sustainable development at a discursive level, with little material or empirical evidence to validate such claims. In turn, it remains important to consider that the most sustainable approach to sports mega-events is likely not to have them at all (see Boykoff, 2014; Boykoff & Mascarenhas, 2016).

Sport and/as eco-justice

The third theme to emerge from the symposium was that of eco-justice. While there are large-scale, macro interventions to be made within the sporting industry, there is also a sense that sport can have environmental benefits at a social, community and/or personal level. In education, for instance, there is a growing body of literature concerning the impact of time spent in natural and outdoor environments upon multiple aspects of health and social development (see Pryor, Carpenter, & Townsend, 2005; Warren, 2005). Outdoor and environmental education can also lead to both positive youth development and sustainability, particularly for communities and individuals which place a sociocultural emphasis on the natural environment. The social and cultural dimensions of sustainable environmental relations are also of particular importance to many Indigenous communities and territories (Adger et al., 2012; Turner & Clifton, 2009). In this regard, there is a growing recognition that Indigenous knowledge, pedagogies and methodologies may contribute to improved and more sustainable relations, an important contribution given that climate change often disproportionately affects Indigenous territories and communities (e.g. Tsosie, 2007).

Indeed, at the intersection of mainstream pedagogy and Indigenous knowledge are examples of environmental education programs that attempt to incorporate Indigenous epistemologies and ontologies and mobilize decolonizing methodologies towards a more holistic understanding of environmental education, psychosocial development, land-based practices and stewardship. For instance, "two-eyed seeing" has been suggested as a lens through which to promote a socio-ecological approach to health and physical activity (Lavallée & Lévesque, 2013), while oral history and other storytelling methods may help to reconstruct or clarify the "fragmented knowledges" that sustain environmental degradation (e.g. Carter et al., 2014; Castleden et al., 2013). There is therefore a clear opportunity for sport scholars to engage in critical, decolonizing scholarship that considers the political economy and power relations at the nexus of land-use, environmental stewardship and culturally determined recreation.

With that said, concerns about sport and eco-justice present both opportunities and challenges because they also extend into other areas of socio-political life, such as the gendered and classed dynamics of sustainable development. Although the scholarly analysis of gender and climate change is growing, particularly concerning the experiences of women in the global South and the links between environmental issues and gendered-based violence (Arora-Jonsson, 2011; MacGregor, 2017; Simon-Kumar et al., 2018; Stock, 2015), this work remains in its relative infancy with respect to the study of sport and SDP. Indeed, post-colonial feminist

approaches, which are arguably still subjugated within SDP research overall, are crucially important when considering sport and the environment
because women disproportionally bear the brunt of environmental degradation while being simultaneously framed as the world's saviors from climate change (see Leach, 2007).

Relations of social class and race also play out within such dynamics
and structures. Citizens of relatively poor countries, and non-owning classes in any country, continue to face the greatest threat from climate
change while also being least afforded representation in the political and
economic processes that govern and/or exacerbate environmental degradation. Put more simply, poor people are more likely to live amidst pollution (e.g. near airports; with higher emissions from cars, industrial
pollution etc.), while large facilities – like sports stadiums – are still often
built in poor or gentrifying neighborhoods, leading to displacement and
forms of environmental racism (Brouwer et al., 2007; Grano & Zagacki,
2011; Sze, 2009). Overall, the point of connecting sport to eco-justice is to
appreciate (and insist) that sport and/as environmental sustainability is
firmly connected to relations of power, both historical and contemporary.
Thus, analyses of sport and the environment may need to extend beyond
the relatively instrumental questions of infrastructure, policy and even
technology, and embrace the sociological dimensions of sport and its complicity in environmental (in)justice.

Conclusion and overview of chapters

Ultimately, the aim of the *Sport and Sustainable Development: Setting
a Research Agenda* symposium was to identify both key themes and new
areas of research in the field(s) of sport, development and environmental
sustainability. Clearly, significant and ongoing work remains to be done in
this regard. Avenues still to be explored include questioning whether and
how sport and its stakeholders might take on a leadership role in shaping
sustainable development strategies, and/or mobilizing sport towards
achieving sustainability and the SDGs. To date, the role allocated to sport
has arguably been a passive one, relying upon vague discourses of sport's
transformative potential and ability to empower and raise awareness.
Questions therefore also remain as to how sport can be used to drive
greater environmental accountability and how to ensure that sustainability
policies are not only promised, but met. What is the role for individuals
and sports fans in this regard? How might sports persons and advocates
take responsibility to reduce their own carbon footprints and be responsible consumers? In regard to social justice, if we are to take seriously the
notion of decolonization amidst sustainability, how might we (continue to)
decolonize leisure and sport? What are the relationships and tensions in
having for-profit sporting organizations leading sustainable development

strategies? If patterns of production and consumption are not changed, how will sustainability goals be met? And how do issues of human rights in sport fit in to notions of sustainability?

These are admittedly massive questions, with few easy answers. The chapters that follow attend to (though do not completely answer) these questions, using various approaches, locales, cases and contexts, as well as a variety of theories, methods and sub-disciplines from across the social sciences. The chapters are organized according to the three themes discussed above.

The first theme addresses various aspects of the role that sport can or might play in supporting and achieving sustainable development, as well as the various tools available to scholars for studying such issues. In Chapter 2, Larry Swatuk provides an overview of the Sustainable Development Goals and Agenda 2030 and uses this as a backdrop to consider what sport in relation to environmental sustainability might (and should) look like in practice. He argues that SDG 12 – Responsible Consumption and Production should be the focus of sport and sustainability efforts because it encourages critical consideration of sport both within capitalist production and lifestyle choices. Ultimately, he argues that the SDGs provide a useful, though imperfect, framework through which to inspire and reform sport towards greater sustainability. In Chapter 3, Brad Millington and Brian Wilson use the sport of golf and its associated industry as a case through which to explore the politics of sustainable development. They argue for the need to radically rethink golf (and sport more broadly) as it relates to the environment and sustainability, as well as the ways in which such issues connect to international development. Ultimately, they advocate for a "darker green" approach to sport, one that recognizes the natural environment for its inherent value, not its relative worth. Such an approach would see decisions about sport (such as where or whether to build new facilities) made in and through frameworks that extend beyond the concerns of corporate environmentalism. In Chapter 4, Kyle Bunds considers the politics of philanthropy and education at the nexus of water security, international development and sport. Based on his research into the use of sport for international development fundraising, he reflects on the ways in which charities, particularly those focused on water scarcity and infrastructure, use physical activity to encourage donations among global North citizens. Given the limits of this approach and the importance of environmental education, he discusses the possibilities of using immersive education technologies as a way to encourage more genuine thinking and learning about the importance and challenge of environmental sustainability. The theme concludes with Chapter 5, by Devra Waldman and Gavin Weedon, who take a different, though ultimately compatible, approach by analyzing the creation of branded, gated communities in major metropolitan regions of India that utilize iconic elements of England's Marylebone Cricket Club and Lord's Cricket Ground. Through a post-colonial analysis, they explore the "imagineering" at work in such efforts, particularly the purifying of nature through urban, sport-related development. They conclude that studying and researching sport and environmental sustainability on an international scale calls for an approach that recognizes the "denied dependencies" between sustainability, development and other logics of exclusion and domination.

The second theme of the book is covered by four chapters that explore various aspects, roles and impacts of INGOs and sports mega-events on sustainable development. In Chapter 6, John Karamichas examines the issue of environmental sustainability as it relates to the hosting of the Olympic Games, and particularly within the context of the International Olympic Committee's recent reforms entitled *Agenda 2020*. He argues that Agenda 2020 should be understood, at least in part, as the IOC's response to broader critiques of the Olympic Movement and that the Agenda is changing the ways in which sustainability is understood and practiced within the planning of future games. However, through an analysis of Olympic Games impact reports, he also argues that, in both policy and practice, Agenda 2020 falls short of guaranteeing the environmental sustainability of hosting the Olympics. The ongoing challenge, then, is to establish new standards that might move Olympic hosting closer to a truly sustainable model. In Chapter 7, Kyoung-yim Kim offers a case study of the 2018 PyeongChang Winter Olympics and examines two key environmental themes – Low Carbon Green Olympics and Stewardship of Nature – within the Korean government's sustainability mandate for hosting the Games. Through this case, she compares the promises of environmentally sustainable development through Olympic hosting to the international climate policy agenda. She argues that despite encouraging positive changes to Korean environmental policy, the PyeongChang Olympics were marred by poor record keeping regarding sustainability efforts, disagreements among key stakeholders, and negative environmental consequences of building for the Games. Her analysis offers further evidence of the limits of ecological modernization and support for approaching sustainability based on notions of justice. In Chapter 8, Carolyn Prouse examines the *Aedes aegypti* mosquito and the Zika virus in the context of the Rio 2016 Olympics/Paralympics. She does so as a way of exploring how "the non-human world" takes shape through sports mega-events, and in so doing challenges the strict binary between the social and the natural in relation to sport. As an alternative, Prouse considers how the social and natural worlds are co-constituted through sports mega-events. Drawing on urban political ecology, she argues that centering sports mega-events as socio-natural reconfigurations serves to illuminate their urban dynamics in critically informed ways, particularly in relation to notions and questions of sustainability, ultimately questioning the idea of sustainability as simply the protection of an externalized nature. And in Chapter 9, Mitu Sengupta offers a comparative analysis of the 1982 Asian Games and the 2010 Commonwealth Games, both hosted in Delhi, India. Using the notion of competitive nationalism, she illustrates how the hosting of these two events proceeded from, and largely confirmed, an approach to development based on modernization and the need for India to "catch up" to the so-called developed world. This approach resulted in not only a growth in inequality – in material, discursive and spatial terms – but also a fundamental resistance to issues of environmental sustainability.

The third and final theme of the book takes up the issues of eco-justice and decolonization more specifically, particularly through an educational lens. In Chapter 10, Tanya Halsall and Tanya Forneris examine and consider opportunities within Sport for Development (SfD) to promote young people's connection to nature and the environment in order that youth might be encouraged and better prepared to contribute towards environmental sustainability. Using ideas from outdoor recreation programming and ecological systems theory, they argue that SfD programming can deliver benefits for youth, community and the environment by considering youth development within multiple levels of influence. They further argue that such approaches and insights may be relevant within an Indigenous context, especially if "two-eyed seeing" is employed to encourage both Indigenous and non-Indigenous perspectives, and to inspire the integration of strengths from each respective cultural viewpoint. In Chapter 11, Dan Henhawk and Richard Norman remind scholars of sport that for many Indigenous people, sport is deeply rooted in racist and colonial notions of Indigenous inferiority and European superiority. It is for these reasons that some contemporary SfD programs aimed at Indigenous communities are met with skepticism and/or resistance. They use this as a point of departure to privilege Indigenous perspectives on sustainable development through sport, and do so in part by exposing the tensions within modernist notions of development as progress. Overall, their chapter serves as an important reminder of the need for Indigenous voices, cultures and ways of thinking within the still burgeoning field of SfD and the issue of sustainability in and through sport. And in Chapter 12, Vicky Paraschak and Michael Heine examine the possibilities afforded when Indigenous land-based practices are included within the paradigm of SfD, especially within the context of Canada's ongoing process of reconciliation within Indigenous peoples. In building this argument, Paraschak and Heine focus less on traditional or dominant notions of sport and more on the cultures and traditions of Indigenous physical activity that emerged from connections between Indigenous people and the land. In attempting to "recover" sport in this way, they suggest that a "strengths and hope" perspective (as opposed to a development approach), one that is informed by notions of land and sustainability, offers an important augmentation to the current SfD paradigm because it repositions Indigenous cultures as assets rather than objects of and for the developmental benefits of sport.

Note

1 In May 2017, the UN announced the closure of the UNOSDP, largely in deference to the efforts of the International Olympic Committee, and in order to avoid any duplication between the two organizations with respect to SDP (Wickstrom, 2017). Despite this significant shift, the SDP sector remains firmly established and recognized.

References

Adger, W. N., Barnett, J., Brown, K., Marshall, N., & O'Brien, K. (2012). Cultural dimensions of climate change impacts and adaptation. *Nature Climate Change*, 3(2), 112–117. 10.1038/nclimate1666.

Adger, W. N., Huq, S., Brown, K., Conway, D., & Hulme, M. (2003). Adaptation to climate change in the developing world. *Progress in Development Studies*, 3(3), 179–195. 10.1191/1464993403ps060oa.

Arora-Jonsson, S. (2011). Virtue and vulnerability: Discourses on women, gender and climate change. *Global Environmental Change*, 21(2), 744–751. 10.1016/J. GLOENVCHA.2011.01.005.

Bale, J. (1994). *Landscapes of modern sport*. London: Leicester University Press.

Black, D., & Van Der Westhuizen, J. (2004). The allure of global games for "semi-peripheral" polities and spaces: A research agenda. *Third World Quarterly*, 25(7), 1195–1214. 10.1080/014365904200281221.

Bolsmann, C., & Brewster, K. (2009). Mexico 1968 and South Africa 2010: Development, leadership and legacies. *Sport in Society*, 12(10), 1284–1298.

Boykoff, J. (2014). *Celebration capitalism and the Olympic Games*. Abingdon, UK: Routledge.

Boykoff, J., & Mascarenhas, G. (2016). The Olympics, sustainability, and greenwashing: The Rio 2016 summer Games. *Capitalism Nature Socialism*, 27(2), 1–11. 10.1080/10455752.2016.1179473.

Brewster, C. (2010). Changing impressions of Mexico for the 1968 Games. *Bulletin of Latin American Research*, 29(1), 23–45.

Briassoulis, H. (2010). "Sorry golfers, this is not your spot!": Exploring public opposition to golf development. *Journal of Sport & Social Issues*, 34(3), 288–311. 10.1177/0193723510377314.

Brouwer, R., Akter, S., Brander, L., & Haque, E. (2007). Socioeconomic vulnerability and adaptation to environmental risk: A case study of climate change and flooding in Bangladesh. *Risk Analysis*, 27(2), 313–326. 10.1111/j.1539-6924.2007.00884.x.

Cameron, J., & Haanstra, A. (2008). Development made sexy: How it happened and what it means. *Third World Quarterly*, 29(8), 1475–1489. 10.1080/01436590802528564.

Cantelon, H., & Letters, M. (2000). The making of the IOC environmental policy as the third dimension of the Olympic Movement. *International Review for the Sociology of Sport*, 35(3), 294–308. 10.1177/101269000035003004.

Carter, C., Lapum, J. L., Lavallée, L. F., & Martin, L. S. (2014). Explicating positionality: A journey of dialogical and reflexive storytelling. *International Journal of Qualitative Methods*, 13, 362–376.

Castleden, H., Daley, K., Sloan Morgan, V., & Sylvestre, P. (2013). Settlers unsettled: Using field schools and digital stories to transform geographies of ignorance about Indigenous peoples in Canada. *Journal of Geography in Higher Education*, 37(4), 487–499. 10.1080/03098265.2013.796352.

Chernushenko, D. (1994). *Greening our games: Running sports events and facilities that won't cost the Earth*. Ottawa, Canada: Centurion.

Coalter, F. (2007). *A wider role for sport: Who's keeping score?* Abingdon, UK: Routledge.

Cooper, A. F. (2007). *Celebrity diplomacy and the G8: Bono and Bob as legitimate international actors*. Waterloo, Canada: Centre for International Governance Innovation, Working Paper 29.

Cornelissen, S. (2010). The geopolitics of global aspiration: Sport mega-events and emerging powers. *International Journal of the History of Sport*, 27(16–18), 3008–3025. 10.1080/09523367.2010.508306.

Cornelissen, S. (2011). More than a sporting chance? Appraising the sport for development legacy of the 2010 FIFA World Cup. *Third World Quarterly*, 32(3), 503–529. 10.1080/01436597.2011.573943.

Darnell, S. C. (2007). Playing with race: Right to Play and the production of whiteness in "development through sport". *Sport in Society*, 10(4), 560–579. 10.1080/17430430701388756.

Darnell, S. C. (2012). Olympism in action, Olympic hosting and the politics of "sport for development and peace": Investigating the development discourses of Rio 2016. *Sport in Society*, 15(6), 869–887. 10.1080/17430437.2012.708288.

Donnelly, P. (2011). From war without weapons to sport for development and peace: The Janus-face of sport. *SAIS Review*, 31(1), 65–76. 10.1353/sais.2011.0015.

Gaffney, C. (2010). Mega-events and socio-spatial dynamics in Rio de Janeiro, 1919–2016. *Journal of Latin American Geography*, 9(1), 7–29. 10.1353/lag.0.0068.

Gaffney, C. (2013). Between discourse and reality: The un-sustainability of mega-event planning. *Sustainability*, 5, 3926–3940. 10.3390/su5093926.

Giulianotti, R. (2012). The sport, development and peace sector: A model of four social policy domains. *Brown Journal of World Affairs*, 18(2), 279–293. 10.1017/S0047279410000930.

Grano, D. A., & Zagacki, K. S. (2011). Cleansing the superdome: The paradox of purity and post-Katrina guilt. *Quarterly Journal of Speech*, 97(2), 201–223. 10.1080/00335630.2011.560175.

Grant, J. (2015). Live Aid/8: Perpetuating the superiority myth. *Critical Arts: South-North Cultural Media Studies*, 29(3), 310–326. 10.1080/02560046.2015.1059547.

Hartmann, D., & Kwauk, C. (2011). Sport and development: An overview, critique, and reconstruction. *Journal of Sport & Social Issues*, 35(3), 284–305. 10.1177/0193723511416986.

Hayes, G., & Karamichas, J. (Eds.). (2012). *Olympics games, mega-events, and civil societies: Globalization, environment, resistance.* New York: Palgrave Macmillan.

Hayhurst, L. (2009). The power to shape policy: Charting sport for development and peace policy discourses. *International Journal of Sport Policy and Politics, 1*(2), 203–227. 10.1080/19406940902950739.

Hayhurst, L. (2011). Corporatising sport, gender and development: Postcolonial IR feminisms, transnational private governance and global corporate social engagement. *Third World Quarterly, 32*(3), 531–549. 10.1080/01436597.2011.573944.

Hayhurst, L., & Szto, C. (2016). Corporatizating activism through sport-focused social justice? Investigating Nike's corporate responsibility initiatives in Sport for Development and Peace. *Journal of Sport and Social Issues, 40*(6), 522–544. 10.1177/0193723516655579.

IOC. (2014, June). IOC plays a key role at first ever UNEA. Retrieved February 25, 2019, from www.olympic.org/news/ioc-plays-key-role-at-first-ever-united-nations-environment-assembly-unea

Jarvie, G. (2012). *Sport, culture and society: An introduction.* Abingdon, UK: Routledge. 10.4324/9780203883808.

Karamichas, J. (2013). *The Olympic Games and the environment.* New York: Palgrave Macmillan.

Kirsch, S. (2010). Sustainable mining. *Dialectical Anthropology, 34*(1), 87–93. 10.1007/s.

Lavallée, L., & Lévesque, L. (2013). Two-eyed seeing: Physical activity, sport, and recreation promotion in Indigenous communities. In J. Forsyth & A. Giles (Eds.), *Aboriginal peoples and sport in Canada: Historical foundations and contemporary issues* (pp. 206–228). Vancouver, Canada: UBC Press.

Leach, M. (2007). Earth mother myths and other ecofeminist fables: How a strategic notion rose and fell. *Development and Change, 38*(1), 67–85.

Lenskyj, H. (1998). Sport and corporate environmentalism. *International Review for the Sociology of Sport, 33*(4), 341–354.

Levermore, R. (2011). The paucity of, and dilemma in, evaluating corporate social responsibility for development through sport. *Third World Quarterly, 32*(3), 551–569. 10.1080/01436597.2011.573945.

MacGregor, S. (Ed.). (2017). *Routledge handbook of gender and environment.* Abingdon, UK: Routledge. 10.4324/9781315886572.

Miller, T. (2016). Greenwashed sports and environmental activism: Formula 1 and FIFA. *Environmental Communication, 10*(6), 719–733. 10.1080/17524032.2015.1127850.

Miller, T. (2018). *Greenwashing sport.* New York: Taylor & Francis.

Millington, B., & Wilson, B. (2014). An unexceptional exception: Golf, pesticides, and environmental regulation in Canada. *International Review for the Sociology of Sport, 51*(4), 1–22. 10.1177/1012690214526878.

Millington, B., & Wilson, B. (2015). Golf and the environmental politics of modernization. *Geoforum, 66*, 37–40. 10.1016/j.geoforum.2015.08.013.

Millington, R., & Darnell, S. C. (2014). Constructing and contesting the Olympics online: The internet, Rio 2016 and the politics of Brazilian development. *International Review for the Sociology of Sport, 49*(2), 190–210. 10.1177/1012690212455374.

Millington, R., Darnell, S. C., & Millington, B. (2018). Ecological modernization and the Olympics: The case of golf and Rio's "green" Games. *Sociology of Sport Journal*, *35*(1), 8–16. 10.1123/ssj.2016-0131.

Mol, A. P. J. (2010). Sustainability as global attractor: The greening of the 2008 Beijing Olympics. *Global Networks*, *10*, 510–528. 10.1111/j.1471-0374.2010.00289.x.

Neo, H. (2010). Unravelling the post-politics of golf course provision in Singapore. *Journal of Sport and Social Issues*, *34*(3), 272–287.

Nicholls, S., Giles, A. R., & Sethna, C. (2011). Perpetuating the "lack of evidence" discourse in sport for development: Privileged voices, unheard stories and subjugated knowledge. *International Review for the Sociology of Sport*, *46*(3), 249–264. 10.1177/1012690210378273.

Pryor, A., Carpenter, C., & Townsend, M. (2005). Outdoor education and bush adventure therapy: A socio-ecological approach to health and wellbeing. *Journal of Outdoor and Environmental Education*, *9*(1), 3–13.

Qing, L. (2010). Encoding the Olympics – visual hegemony? Discussion and interpretation on intercultural communication in the Beijing Olympic Games. *International Journal of the History of Sport*, *27*(9), 1824–1872. 10.1080/09523367.2010.481136.

Richey, L. A., & Ponte, S. (2008). Better (Red)^{TM} than dead? Celebrities, consumption and international aid. *Third World Quarterly*, *29*(4), 711–729. 10.1080/01436590802052649.

Simon-Kumar, R., MacBride-Stewart, S., Baker, S., & Saxena, L. P. (2018). Towards north-south interconnectedness: A critique of gender dualities in sustainable development, the environment and women's health. *Gender, Work & Organization*, *25*(3), 246–263. 10.1111/gwao.12193.

Stock, A. (2015). Beijing, gender and environment: Challenges for ecological sustainability, development and justice? *IDS Bulletin*, *46*(4), 54–58.

Sze, J. (2009). Sports and environmental justice: "Games" of race, place, nostalgia, and power in neoliberal New York City. *Journal of Sport and Social Issues*, *33*(2), 111–129.

Trendafilova, S., Babiak, K., & Heinze, K. (2013). Corporate social responsibility and environmental sustainability: Why professional sport is greening the playing field. *Sport Management Review*, *16*(3), 298–313. 10.1016/j.smr.2012.12.006.

Tsosie, R. (2007). Indigenous People and environmental justice: The impact of climate change. *University of Colorado Law Review*, *78*(1), 1625–1678. 10.3868/s050-004-015-0003-8.

Turner, N. J., & Clifton, H. (2009). "It's so different today": Climate change and indigenous lifeways in British Columbia, Canada. *Global Environmental Change*, *19*(2), 180–190. 10.1016/j.gloenvcha.2009.01.005.

United Nations. (1972). *Declaration of the United Nations Conference on the Human Environment*. Stockholm, Sweden: UN General Assembly.

United Nations. (2016). *Sport and the Sustainable Development Goals: An overview outlining the contribution of sport to the SDGs*. New York: UN Office on Sport for Development and Peace.

Warren, K. (2005). A path worth taking: The development of social justice in outdoor experiential education. *Equity & Excellence in Education*, *38*(1), 89–99.

WCED. (1987). *Report of the World Commission on Environment and Development: Our common future*. New York: United Nations.

Wickstrom, M. (2017). UN Secretary-General closes UNOSDP. Retrieved May 11, 2017, from www.playthegame.org/news/news-articles/2017/0309_un-secretary-general-closes-unosdp/

Wilson, B., & Millington, B. (2013). Sport, ecological modernization, and the environment. In D. L. Andrews & B. Carrington (Eds.), *A companion to sport* (pp. 130–142). Hoboken, NJ: Blackwell.

Zhang, T., & Silk, M. L. (2006). Recentering Beijing: Sport, space, and subjectivities. *Sociology of Sport Journal*, *23*, 438–459.

Chapter 2

Add sport and stir?

The SDGs and sport-environment-development

Larry A. Swatuk

Introduction

In September 2015, the United Nations General Assembly adopted the Sustainable Development Goals (SDGs) as the macro-operating framework for development. The SDGs follow on from the Millennium Development Goals (MDGs) which ran for 15 years, from 2001 to 2015. The MDGs were adopted by 160 world leaders at the United Nations in September 2000. The eight goals overwhelmingly focused on socio-economic problems in the Global South. In the Foreword to the 2015 MDG Report, Ban Ki-Moon, the Secretary-General of the United Nations, claimed that 'The MDGs helped to lift more than one billion people out of extreme poverty, to make inroads against hunger, to enable more girls to attend school than ever before and to protect our planet.' There have been many criticisms of the MDG exercise (Hickel, 2016), an important one of which is that economic growth in India and China ensured progress was made across the goals irrespective of actions taken in direct support of the MDGs (see www.xinhuanet.com/english/2015-09/26/c_134661386.htm). In a recent study by McArthur and Rasmussen (2018), the authors show that the MDGs helped accelerate gains made in low-income countries (excluding India), particularly in areas of acute need admitting of specific and measurable interventions (i.e. child mortality, maternal mortality, water and sanitation). They were less effective, however, in accelerating progress across the same four indicators in middle-income countries (excluding China). Granted, attributing change directly to any one or a set of factors within a complex system is a dangerous game to play. What is certain, however, is that the MDGs provided a common language and organizational framework for thinking about and addressing broadly accepted 'global goals'.

The SDGs, first mooted at the *United Nations Rio +20 Conference on Sustainable Development* held at Rio de Janeiro in 2012 (UN, 2012: 63–64), are set to run until 2030 and provide the same kind of discursive framework within which to determine, develop, deliver, monitor and assess action toward the realization of what is called 'Agenda 2030' (cf. Briant Carant, 2017). A significant difference between the SDGs and the MDGs is the balanced focus across Global North and Global South. Whereas SDGs 1–6 and 17 directly echo the eight MDGs, SDGs 7–16 ask important questions not only about the consequences of under-consumption and under-development primarily in the Global South, but – and perhaps more importantly – of over-consumption and over-development primarily in the Global North. A key difference between the MDGs and the SDGs is the latter's more explicit engagement with questions of environmental sustainability, which goes far beyond an isolated goal and toward integration across all 17 goals.

Further, unlike the MDGs, which presented an orthodox developed–developing/helper–helped depiction of the global terrain for 'international' development, the SDGs place a good deal of emphasis on integration and inter-connection, not only across goals (e.g. water, energy, food) and within goals (e.g. production and consumption, inequality, climate change), but across and among states and societies.

This chapter focuses on ways and means of embedding efforts toward the realization of Agenda 2030 within sport, broadly defined. It argues that far from being an 'add-on', sport can have a central impact on Agenda 2030, particularly through focused and sustained engagement with SDG 12, *Responsible Consumption and Production*. Meaningful and useful engagement can take place almost anywhere across the SDG landscape of goals, targets and indicators. However, since sport is most centrally embedded in society through the production of things – from clothing and footwear to infrastructure and events of all shapes and sizes – it is upon this that an Agenda 2030-oriented approach should focus. In support of this argument, the chapter proceeds as follows. The next section describes the SDGs, what are they and what they are meant to do. Following this, the chapter turns to a discussion of sport in the SDGs from two perspectives: an orthodox approach which is happy to 'contribute'; and a more innovative approach which highlights sport's ability to lead. As mentioned above, the focus here is on SDG 12. The chapter offers several examples from the corporate world in support of its main argument. It concludes with a call to action, recognizing that our choices determine our chances of achieving some, all or none of the SDGs.

A 'primer' on the SDGs

The SDGs are often called 'the global goals' or 'the people's goals' (OECD, 2016). According to Osborn et al. (2015: 2) the SDGs,

> are intended to be universal in the sense of embodying a universally shared common global vision of progress towards a safe, just and

sustainable space for all human beings to thrive on the planet. They reflect the moral principles that no-one and no country should be left behind, and that everyone and every country should be regarded as having a common responsibility for playing their part in delivering the global vision.

The goals are as follows:

- SDG 1: End poverty in all its forms everywhere;
- SDG 2: End hunger, achieve food security and improved nutrition and promote sustainable agriculture;
- SDG 3: Ensure healthy lives and promote well-being for all at all ages;
- SDG 4: Ensure inclusive and equitable education and promote lifelong learning opportunities for all;
- SDG 5: Achieve gender equality and empower all women and girls;
- SDG 6: Ensure availability and sustainable management of water and sanitation for all;
- SDG 7: Ensure access to affordable, reliable, sustainable and modern energy for all;
- SDG 8: Promote sustained, inclusive and sustainable economic growth, full and productive employment and decent work for all;
- SDG 9: Build resilient infrastructure, promote inclusive and sustainable industrialization and foster innovation;
- SDG 10: Reduce inequality within and among countries;
- SDG 11: Make cities and human settlements inclusive, safe, resilient and sustainable;
- SDG 12: Ensure sustainable consumption and production patterns;
- SDG 13: Take urgent action to combat climate change and its impacts;
- SDG 14: Conserve and sustainably use the oceans, seas and marine resources for sustainable development;
- SDG 15: Protect, restore and promote sustainable use of terrestrial ecosystems, sustainably manage forests, combat desertification, and halt and reverse land degradation and halt biodiversity loss;
- SDG 16: Promote peaceful and inclusive societies for sustainable development, provide access to justice for all and build effective, accountable and inclusive institutions at all levels;
- SDG 17: Strengthen the means of implementation and revitalize the Global Partnership for Sustainable Development.

While responsibility for achieving the SDGs lies primarily with states and their governments, a number of collaborative entities and spaces have been established. For example, the Sustainable Development Solutions Network (SDSN) was created in 2012 and operates under the auspices of the UN Secretary-General. The SDSN 'mobilizes global scientific and technological expertise to promote practical solutions for sustainable development' (http://unsdsn.org/about-us/vision-and-organization/). The SDSN also establishes national and regional networks of universities, research centres and other forms of knowledge mobilization organizations to facilitate localization of the SDGs and their implementation (http://net works.unsdsn.org/). Currently, there are more than 800 members in the network.

The Organisation for Economic Co-operation and Development (OECD) has developed an Action Plan in support of implementation of the SDGs by and among its members – that include mostly high-incomes countries – and their partners (www.oecd.org/dac/Better%20Policies%20for%202030.pdf). At regional level, the European Union has developed a consensus document in support of Agenda 2030 (https://ec.europa.eu/europeaid/new-european-consensus-development-our-world-our-dignity-our-future_en). In relation to Africa, the African Union established its own Agenda 2063, which predates Agenda 2030, but is entirely compatible with it (https://au.int/en/ea/statistics/ a2063sdgs). In turn, all governments around the world have dedicated offices and personnel in support of the SDGs. For example, in its 2018 budget the Government of Canada 'allocated new funds to establish an SDG Unit to ensure effective Agenda 2030 coordination across federal departments and agencies and with Canadian stakeholders, and to track Canada's progress on the SDGs' (GoC, 2018: 7).

Despite all of this coordinated action, evidence shows that awareness of the SDGs is limited, though increasing. An OECD DevComms (2017) report cites Eurobarometer survey data showing awareness of the SDGs increasing across 16 European countries from 36% to 41% between 2016 and 2017, and knowledge increasing from a meagre 10% to 12%. Survey data also shows major disparities between countries, and 'AIESEC's Youth Speak report (2016) suggests that young people have a higher level of SDG awareness than average [45%], a finding that is generally replicated in demographic analyses of other surveys' (OECD DevComms, 2017).

In relation to performance, according to the SDG Index and Dash-boards Report 2018, most of the Global North scores in the 70–80 range of achievement of the sustainable development goals (where 100 equals perfect achievement of the global goals; see https://dashboards .sdgindex.org/). Only seven countries score 80 or higher (Austria 80; Norway and France 81.2; Germany 82.3; Finland 83; Denmark 84.6; Sweden 85). A wide range of countries score 70 or higher – from Azerbaijan and Uzbekistan to Uruguay, Australia, all of Eastern Europe, the USA and Canada. Great swaths of Latin America, Sub-Saharan Africa and South Asia score considerably lower. Given that there are 17 goals with 169 targets and roughly 230 indicators, differences within and across countries sharing similar scores can be significant.[1]

Sport and the SDGs

As with the MDGs, there has been much speculation regarding the role of sport in the SDGs. Below are descriptions of various ways in which sport has been imagined in, and has engaged with, the SDGs. The SDGs provide an important opportunity for those engaged with and in sport to make meaningful contributions to Agenda 2030. In order to realize this opportunity, however, it is important to reflect critically on the place of sport in society and to see it as generally reflective of dominant social forms and practices. Put differently, if sport is a reflection of society as currently constituted, it often functions to deepen dominant social forms and practices, rather than challenge them (Black, 2017; Black and Peacock, 2011; Black and Northam, 2017). Rare is the occasion where sport departs from dominant discourses and practices and seeks to shape society in a direct and meaningful way. Similarly, McArthur and Rasmussen (2018) argue that the least advancement over the course of the MDGs was made in terms of environmental sustainability. This should not be surprising, since rapid industrial growth in India and China led to a decades-long bonanza for extractive industries. Achieving a halving of the number of people in absolute poverty came at a very high cost to the environment. It is because of this that SDG 12 is at the heart of Agenda 2030. Without dramatically improved approaches to, and practices of, industrial production, none of the other goals are 'sustainable' (Millennium Ecosystems Assessment, 2005). It is also through SDG 12 that sport has the greatest opportunity to contribute to the environment and development, to be a leader not a follower, or to determine development and not just be a part of it. At the same time, there are various pathways for sport to take in enabling and fostering sustainable development.

SDGs and sport: the orthodox view

In the UN Declaration of the 2030 Agenda for Sustainable Development, sport is described as 'an important enabler of sustainable development', partly because 'sport has proven to be a cost effective and flexible tool in promoting peace and development objectives'. Put differently, sport offers a 'soft' pathway for important actions such as trust-building, social interaction, educational opportunities and health promotion that are often difficult to address directly in a conflictual and divided world.

UN Resolution A/69/LS, adopted 16 October 2014, 'encourages Member States to give sport due consideration in the context of the post-2015 development agenda'. The recently closed UN Office on Sport for Development and Peace also argued that the SDGs provide a number of entry points for sport. For example:

- Sport can develop transferable skills and toolkits which play a key role in a self-reliant and sustainable life and lead to income-generating activities and economic participation. It can advocate for ending poverty as well as generate funds and facilitate partnerships for this goal [SDG 1].
- Sport can raise awareness of sustainable food sourcing, food security, healthy nutrition and sustainable agriculture. Sport organizations can set an example by sourcing food from responsible producers and tackling food waste impact. Sport-based educational initiatives can aim at changing behaviours towards a sustainable future [SDG 2].

Beyond the UN system, inter-governmental organizations such as the Commonwealth and the Association of Southeast Asian Nations (ASEAN) have speculated on the role of sport in realizing the SDGs. Ahead of the September 2015 launch of the SDGs, the Commonwealth urged governments to 'recognize the value of sport in sustainable development' (http://thecom monwealth.org/media/press-release/governments-urged-recognise-value-sport-sustainable-development). For ASEAN member states, 'The use of sport in pursuit of the … SDGs … is broadening, likely due to its versatility and ability to engage diverse segments of society [where] other mediums may fall short' (www.sportanddev.org/en/event/asean-sport-and-sdgs-youth-funshop). In a 2015 report (http://thecommonwealth.org/sites/default/files/inline/CW_SDP_2030%2BAgenda.pdf), the Commonwealth highlighted the ways in which sport can assist in the realization of specific SDGs. For the Commonwealth, SDGs 3 (health), 4 (education), 5 (gender equality), 8 (economic growth), 11 (cities), 16 (peace) and 17 (cross-cutting theme regarding partnerships) offer immediate pathways for sport.

Given the interlinkages across the SDGs, one could argue that progress on SDGs 1 and 2 will positively impact SDG 3, which is focused on health and well-being. For example, the SDGs Index and Dashboard Report for 2018 (https://dashboards.sdgindex.org/) shows that in relation to SDG 2, obesity rates are high and rising in high income countries. Specifically, 29.4% of the population in Canada, 27.8% in the UK, and 36.2% in the USA have a body mass index greater than or equal to 30. This is on par with many low-income countries across Sub-Saharan Africa, where poverty, ill-health (including persistently high tuberculosis and HIV/AIDS infection rates), low life expectancy and high rates of teenage pregnancy (all indicators in SDG 3) are also present. The entry points for sport here are clear.

A pathway for sport in sustainable development has also been laid out by the International Olympic Committee (IOC) through its 2007 *Guide to Sport, Environment and Sustainable Development*, particularly the first two chapters of the document. In its 2016 sustainability strategy, the IOC highlighted sports' linkages to the same array of SDGs as articulated by the

Commonwealth, but went a step further by specifying its own five areas of focus: infrastructure and natural sites; sourcing and resource management; mobility; workforce; and climate. In an undated document, the UN Office on Sport for Development and Peace articulated the myriad ways in which sport can contribute directly to the SDGs (UNOSDP, n.d.). This document begins with a quotation from the UN Political Declaration for the new Agenda:

> Sport is also an important enabler of sustainable development. We recognize the growing contribution of sport to the realization of development and peace in its promotion of tolerance and respect and the contributions it makes to the empowerment of women and young people, individuals and communities as well as to health, education and social inclusion objectives (2030 Agenda for Sustainable Development A/RES/70/1, paragraph 37).
>
> (UNOSDP, n.d.: 2)

The document goes on to argue that the role of sport in the SDGs should aim at 'informing, encouraging and supporting sport's contributions to the SDGs by relevant stakeholders including States, entities of the United Nations system, sport-related organizations, sport federations and associations, foundations, non-governmental organizations, athletes, the media, civil society, academia and the private sector' (UNOSDP, n. d.: 2).

These are inspiring words indeed. However, a critical reading of these various documents finds that the language is more passive than it is active: sport 'can develop', 'can advocate', 'can help', 'can support'. This stands in opposition to sport 'doing', 'ensuring' or 'mandating'. In sum, it seems that many sport organizations are content to play a secondary role in sustainability by promoting, encouraging, motivating, fostering, disseminating and raising awareness, rather than acting in sustainable ways (see Lemke, 2016).

Sport as a 'powerful tool'

While those in support of sport obviously regard it as a 'powerful' and/or 'effective' tool for change across a number of issue areas, just how this power will be brought to bear is not made clear. Sport has a global reach, offers a near universal language and has extraordinary 'convening power' at all scales, regularly demonstrating an ability to rally communities, engage youth, reach vulnerable groups and create shared interests. At the same time, the 'tool' of sport is only as effective as the (segments of) society that wield(s) it. Put differently, sport is a reflection of settled social organization. In terms of scale, it is equal parts global business and local pastime. It is embedded in place, but varies across geographical and cultural landscapes, giving rise to specific production and consumption. As the recent documentary film *The Workers Cup* (see www.theworkerscupfilm .com/) shows in relation to preparations for the 2022 Football World Cup in Qatar, sport across its 'value chain' is infused with class, race, gender, age, ethnicity and religion. As such, sport divides as much as it unifies – not only at and on the football ground, but also across the landscape of production and consumption – and often reifies more than it reforms or revolutionizes. For example, the primary 'winners' of European football are often economic oligarchs such as Chelsea football club's owner Roman Abramovich, as well as the club's primary shareholders, despite its 'progressive' partnerships with global non-governmental organizations such as Right to Play and Plan International. The growth of philanthropy in the global political economy reflects neoliberal globalization's advantages to the 1%, at significant cost to planet and people (Drury, 2014). In almost every way, sport reflects such tendencies, acting as a powerful instrument for reinforcing, deepening, or 'naturalizing' existing structures of wealth and power.

What then, is to be done? Can sport play a central, active, mobilizing, transformative, or even revolutionary role toward the realization of a more environmentally sustainable and socially equitable world like that envisioned by the supporters of the SDGs? In my view, in addition to mainstreaming efforts across all 17 SDGs, sport can make its most meaningful contribution through SDG 12: *Responsible Consumption and Production*. Achieving the goal of responsible consumption and production will likely mean that the other goals will have resolved themselves, for it is how we access, allocate and use the planet's resources that give form and substance to human social relations. If we were to achieve socially equitable, economically efficient and environmentally sustainable use of resources, we would have remade the world. Indeed, to paraphrase the title of the UN's Rio +20 final report, we would have arrived at 'the future we want'. Results from a 2015 stakeholder survey by Osborn et al. identify:

> the goals of sustainable consumption and production (SDG 12), sustainable energy (SDG 7) and combating climate change (SDG 13) as the three most transformational challenges facing developed countries – and those on which the developed world needs to place a strong emphasis for action so as to relieve the overall anthropogenic pressures on the planet and its natural systems.
>
> (Osborn et al., 2015: 2)

This data also shows the importance of 'more sustainable economies and growth pathways, the goal of greater equality, and the goals to achieve better protection of the oceans and of terrestrial ecosystems', thus demonstrating the integrated nature of production and consumption across the global socio-ecological landscape.

The will to lead: sport and SDG 12

Paragraph 61 of the UN's Rio +20 report entitled *The Future We Want* states:

> We recognize that urgent action on unsustainable patterns of production and consumption where they occur remains fundamental in addressing environmental sustainability and promoting conservation and sustainable use of biodiversity and ecosystems, regeneration of natural resources and the promotion of sustained, inclusive and equitable global growth.
>
> (UN, 2012: 16)

The report also places central emphasis on 'the green economy' and the necessity of transitioning away from carbon-centric systems of production and consumption that regard the Earth as a boundless source of materials and/or a sink for human waste. Much has been written about the many ways in which neoliberal globalization has exacerbated such problems, as much as it has helped alleviate poverty in particular parts of 'emerging economies' and rising states (Panitch and Gindin, 2012; Ponting, 2007). Several of these rising states, such as the so-called BRICS – Brazil, Russia, India, China and South Africa – have sought to demonstrate their new-found power in the form of sport, particularly by hosting sports mega-events (Black and Peacock, 2011), suggesting a continuing commitment to 20th-century measures of political power rather than 21st-century notions of sustainable development (Cornelissen, 2011; Swart and Bob, 2004).

It is also no revelation to say that the production structures surrounding sport vary dramatically depending on the activity, from global games to backyard kickball. The ecological footprint of these activities varies widely, as do the politically economic factors underlying them.[2] As suggested above, sports mega-events have ostensibly made much of their potential to follow an SDG-oriented business model, with the IOC choosing to focus on five areas: infrastructure and natural sites; sourcing and resource management; mobility; workforce; and climate (IOC, 2017). Indeed, the IOC has increasingly pressed for 'green games', with the 2000 Sydney Olympics being declared as such. FIFA similarly described the 2006 World Cup in Germany as aiming to score a 'green goal'. In 2007, UEFA created the Sustainable Development Charter, leading to such things as the Euro 2012 *Social Responsibility Report* (UEFA, 2012), wherein benchmarks were set in terms of sustainable and ethical forms of production and consumption. These are noteworthy ways forward in terms of raising the sustainability bar. However, while such high-profile events provide an important opportunity to demonstrate the (dis)integration of sport with sustainable development practices and processes, individuals and organizations need not wait for the next Olympics or World Cup before acting.

Sport is already connected to and embedded across SDG 12 through various products and practices of sport in the global economy, including clothing, footwear, equipment, infrastructure, advertising, gaming, events and communications. In order to lead and shape improved practices of production and consumption, we need to focus on two key sets of questions: what, where and how is sport produced, and what, where and how is sport consumed? Those of us interested in sustainable development must inform appropriate practices and lead the way in terms of making choices. In this view, there are several ways to connect SDG 12 to sport:

• As a matter of course, sports mega-events should routinely adhere to a variety of ethical measures. For example, the International Organisation for Standardization's ISO 14001 (Environmental Management), ISO 26000 (Social Responsibility) and ISO 37101:2016 (Sustainable Development in Communities – Management System for Sustainable Development – Requirements with Guidance for Use).

• We must not simply encourage better practice. It is up to us – as consumers and producers of sport – to show through our everyday practices and choices the ways and means of lightening sport's ecological footprint while pressing for socially just forms of resource use and product development and production. At the micro-scale, we can inform ourselves beyond the price of a product to examine the value chain that helped create it, and to demand from producers more ethical practices across the production landscape. We can also demand from ourselves, our families and our neighbours that lower prices not be the determining factor behind a purchase.

• At the macro-scale of global games, and the meso-scale of professional sports clubs, adherence to Leadership in Energy and Environmental Design certification systems for infrastructure will help avoid the 'white elephants' that emerge everywhere in the lead up to sports mega-events, and ensure sustainable design in the planning and delivery of new or renewed sports venues.

• Sports companies must engage in sustained analysis of their value chains and be transparent in their reporting of such.

• Civil society groups, with support from governments and the private sector, must build consumer-friendly life cycle assessment tools that help consumers to understand the importance of 'cradle-to-cradle' production, and producers to source their raw materials ethically and sustainably.

- Similarly, user-friendly analytical tools such as the 'ecological footprint' should be developed for applicability across the sport landscape. Granted, this is a relatively crude tool but it helps to personalize consumption and, more importantly, act as a contact point between producers and consumers (and all citizens) within and across states.

The key is that sport occurs at every scale – across common spaces and day-to-day forms of play, as well as mega-events. And Sport for Development and Peace advocates have shown the myriad and creative ways in which sport can contribute to social development. There is every reason to believe sport can make similar contributions to the SDGs. To illustrate this point, consider the idea of sport as both a product and as a (global) business. It can then be examined in relation to the Global Reporting Initiative/United Nations Global Compact/World Business Council for Sustainable Development guide for business action on the SDGs, commonly known as the SDG Compass (GRI/UNGC/WBCSD, 2015). The term 'compass' is important here, for it suggests that 'companies can use the SDGs as an overarching framework to shape, steer, communicate and report their strategies, goals and activities, allowing them to capitalize on a range of benefits' (GRI/UNGC/WBCSD, 2015: 4). The compass aims to show businesses how to shift their perspective not only in service of the success of their company, but in service to the planet. In the report's words:

> Today's internally focused approach to goal setting is not enough to address global needs ... By looking at what is needed externally from a global perspective and setting goals accordingly, businesses will bridge the gap between current performance and required performance. The SDGs represent an unprecedented political consensus on what level of progress is desired at the global level.
>
> (GRI/UNGC/WBCSD, 2015: 19)

The SDG Compass (see https://sdgcompass.org/wp-content/uploads/2015/12/019104_SDG_Compass_Guide_2015.pdf) presents a five-step methodology for integrating Agenda 2030 into business practice. Step One centres on 'Understanding the SDGs'. Importantly, and as suggested in the bullet points above, mainstreaming sustainable development in production (and consumption) requires not only understanding the SDGs, but also the wide array of existing normative frameworks, principles and guidelines already in place. In addition to those cited above, the SDG Compass identifies the International Labour Organization Tripartite Declaration of Principles Concerning Multinational Enterprises and Social Policy, the UN

Global Compact Principles and the UN Guiding Principles on Business
and Human Rights. Step Two involves 'Defining Priorities' and suggests as
an important starting point that businesses map the value chain to identify
impact areas. (Figure 2.1 highlights some of the aspects of SDG 12 that
a company should consider along its value chain.)

Step Three is 'Setting Goals', and suggests that companies establish
SDG-sensitive key performance indicators, establish baselines (as points in
time or as periods of time) and set goals that are based on an outside-in
approach rather than an inside-out approach (see Table 2.1). The SDG
Compass developers argue that there is a 'gap' that exists between local/
company/industry-oriented goal setting (i.e. inside-out) and global/social/
integrated needs, interests, pressures and opportunities (i.e. outside-in).

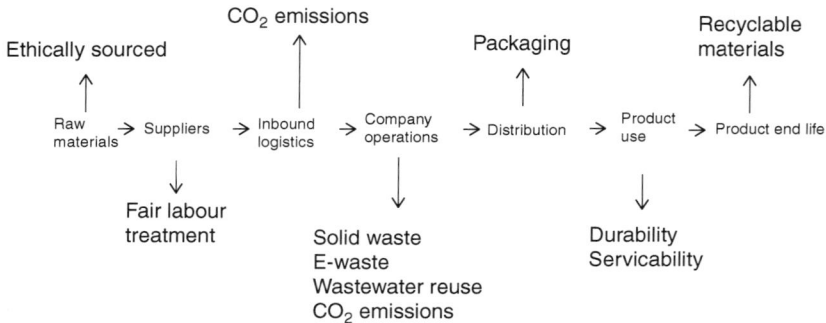

Figure 2.1 Mapping SDG 12 against the value chain

Table 2.1 Adopting an SDG-oriented goal-setting approach

Inside-out approach	Outside-in approach
Combined impact of current business goals (company-focused)	Global and societal needs (SDG-focused)
Set internally	Set based on external societal or global needs
Based on historical data, currents trends and future projections of company's performance	Based on science and external data
Benchmarked against performance and goals of industry peers	Benchmarked against the needs of society that your business can address

Source: Derived from figure in GRI/UNGC/WBCSD (2015: 19).

A company with a broader, more holistic and SDG-sensitive vision may profitably address these without abandoning their 'bottom line'. The SDG Compass highlights several examples of innovative and successful goal-setting approaches by businesses worldwide, including the Science Based Targets Initiative (https://sciencebasedtargets.org/), the WBCSD's Action2020 (http://action2020.org/), the Future-Fit Benchmarks (http://futurefitbusiness.org/) and PivotGoals (www.pivotgoals.com/).

Step Four involves strategies and tactics for 'Integrating' the SDGs into existing business practice and Step Five focuses on 'Reporting and Communicating'.

It is possible to integrate sport into such a framework. The sporting goods and branding corporation Nike offers a useful example. Nike's Sustainable Business Report for fiscal year 2016/17 (Nike, 2017) illustrates very clearly how far the company has come since critics exposed its abusive labour practices and environmental degradation. In terms of raw materials, Nike now reports both the widespread use of recycled materials (through, e.g., Nike Grind), as well as the sourcing of organic cotton. With regard to its suppliers, Nike engages in what Gereffi (1994) identifies as buyer-driven commodity chain production. What this means is that the company is engaged in own-brand manufacturing, but out-sources the actual production to contract factories around the world. In this way, the company takes advantage of such things as highly reduced labour costs, but it makes it vulnerable to inappropriate behaviour. Following extensive criticisms regarding 'sweatshop production' in the 1990s, Nike has taken significant steps to reduce incidences of child labour and poor working conditions in the factories with which it does business. According to its latest report, 532 out of 570 contract factories achieved 'bronze' status, meaning that 91% of its partners were in 'compliance with our code of conduct and show a commitment to lean manufacturing' (Nike, 2017: 42). In terms of inbound logistics, Nike has committed to a 20% reduction in carbon emissions per unit. With regard to company operations, Nike has made considerable progress in terms of reducing its carbon and water footprints. It has set distribution goals such as 10% reductions in shoebox weight per unit, and in terms of end of life of the product, the company engages in an extensive recycling programme.

This brief case study illustrates some of the ways the business world has embraced the challenge of sustainability, often in response to criticisms and exposure levelled by social movements or civil society organizations. But such changes have also come about because the SDG framework provides myriad opportunities for triple bottom line benefits to companies of all shapes and sizes.[3] This new degree of transparency in reporting is unprecedented, providing consumers with an ability to hold producers to existing commitments, pressure them for better practice, support their efforts or boycott them altogether. While not a guarantee against 'green-washing', SDG goals, targets and indicators provide globally agreed benchmarks by which to judge and, if necessary, challenge corporate performance. The wide variety of metrics available in support of SDG implementation also allows for the adaptation of tools such as the SDG Compass for use by other actors, from households to sports clubs to communities. Enabled by internet technology and telecommunications innovations, the SDG framework provides a common language, a shared platform and therefore a unique opportunity for engagement among states, civil society organizations, private sector actors and individuals.

Conclusion

We should not be naïve about the potential for arriving at transformational change in environmental sustainability on a global scale. The tension between profit and planet continues. The rise of the 'one percent' and new nationalist movements indicate a widening gap among people at precisely the time when we need to pull together (Harari, 2018; Ingraham, 2017). As with the MDGs, the SDGs face all sorts of criticism. In Easterly's (2015) words, 'SDG' should stand for 'senseless, dreamy and garbled'. However, in trying to summarize all that needs to be done in service of people and the planet, a degree of vagueness is, in my view, a good thing. The entry points for positive action are almost limitless. Like it or not, Agenda 2030 is the accepted framework for development among the 192 signatories to the document. It appears to be particularly appealing to youth who have grown up in the forbidding shadow of climate change, biodiversity loss and land degradation. Sport, and the places we play, are at the centre of these impacts. In turn, the ways that we play might exacerbate or ameliorate these places and spaces. It is therefore time to inform ourselves and act in support of informed choices in ways that might position sport in the service of environmental sustainability.

Notes

1 For example, in relation to SDG 2, there are eight specific targets and 14 indicators (see https://sustainabledevelopment.un.org/content/documents/11803Official-List-of-Proposed-SDG-Indicators.pdf).
2 For details on the ecological footprint, see Wackernagel and Rees (1996).
3 As highlighted by Swart et al., 'Increasing popular pressure on international sport federations and their commercial partners to demonstrate greater levels of accountability has obliged them to take the issue of the broader developmental significance of their events more seriously' (Swart et al., 2011:420).

References

Black, D. and K. Northam, 2017. Mega-events and 'bottom-up' development: Beyond window dressing? *South African Journal for Research in Sport, Physical Education & Recreation* 39, 1–17.

Black, D. and B. Peacock, 2011. Catching up: Understanding the pursuit of major games by rising developmental states. *International Journal of the History of Sport* 28:16, 2271–2289.

Black, D.R., 2017. The challenges of articulating 'top down' and 'bottom up' development through sport. *Third World Thematics: A TWQ Journal* 2:1, 7–22.

Briant Carant, J., 2017. Unheard voices: A critical discourse analysis of the Millennium Development Goals' evolution into the sustainable development goals. *Third World Quarterly* 38:1, 16–41.

Cornelissen, S., 2011. More than a sporting chance? Appraising the sport for development legacy of the 2010 FIFA World Cup. *Third World Quarterly* 32:3, 503–529.

Drury, I., 2014. The rise of philanthrocapitalism: What passes for progressive city politics today. *Briarpatch*. Available at: https://briarpatchmagazine.com/articles/view/the-rise-of-philanthrocapitalism accessed 11 July 2019.

Easterly, W., 2015. The SDGs should stand for 'senseless, dreamy and garbled'. *Foreign Affairs*, 28 September.

Gereffi, G., 1994. The organization of buyer-driven global commodity chains: How U.S. retailers shape overseas production networks. In: Gereffi, G. and M. Korzeniewicz, eds, *Commodity Chains and Global Capitalism*. New York: Praeger, pp. 95–122.

Government of Canada, 2018. *Canada's Implementation of the 2030 Agenda for Sustainable Development*. Ottawa, Canada: Government of Canada.

GRI/UNGC/WBCSD 2015. *SDG Compass: The Guide for Business Action on the SDGs*. Available at: https://sdgcompass.org/wp-content/uploads/2015/12/019104_SDG_Compass_Guide_2015.pdf accessed 10 September 2018.

Harari, Y.N., 2018. *21 Lessons for the 21st Century*. Harmondsworth, UK: Penguin.

Hickel, J., 2016. The true extent of global poverty and hunger: Questioning the good news narrative of the Millennium Development Goals. *Third World Quarterly* 37:5, 749–767.

Ingraham, C. 2017. The richest 1 percent now owns more of the country's wealth than at any time in the past 50 years. *The Washington Post*, 6 December.

International Olympic Committee (IOC), 2017. *IOC Sustainability Strategy (October)*. Available at: www.olympic.org/~/media/Document%20Library/OlympicOrg/Factsheets-Reference-Documents/Sustainability/IOC-Sustainability-Strategy-Long-version-v12.pdf accessed 28 November 2018.

Lemke, W., 2016. The role of sport in achieving the sustainable development goals. *UN Chronicle* 53(2). Available at: https://unchronicle.un.org/article/role-sport-achieving-sustainable-development-goals accessed 14 July 2018.

McArthur, J. and K. Rasmussen, 2018. Change of pace: Accelerations and advances during the Millennium Development Goal era. *World Development* 105, 132–143.

Millennium Ecosystems Assessment, 2005. *Ecosystems and Human Well-being: Synthesis*. Washington, DC: Island Press.

Nike, 2017. *Maximum Performance Minimum Impact: FY16/17 Sustainable Business Report*. Nike, Inc. Beaverton, OR. Available at: https://s1.q4cdn.com/806093406/files/doc_downloads/2018/SBR-Final-FY16-17.pdf accessed 10 September 2018.

OECD, 2016. *Better Policies for 2030: An OECD Action Plan on the Sustainable Development Goals*. Paris, France: OECD.

OECD DevComms, 2017. *What People Know and Think about the Sustainable Development Goals.* Available at: www.oecd.org/development/pgd/International_Survey_Data_DevCom_June&202017.pdf accessed 10 September 2018.

Osborn, D., A. Cutter and F. Ullah, 2015. *Universal Sustainable Development Goals: Understanding the Transformational Challenge for Developed Countries.* Maidstone, UK: Stakeholder Forum.

Panitch, L. and S. Gindin, 2012. *The Making of Global Capitalism: The Political Economy of the American Empire.* London: Verso.

Ponting, C., 2007. *A New Green History of the World.* Toronto, Canada: Vintage.

Swart, K. and U. Bob, 2004. The seductive discourse of development: The Cape Town 2004 Olympic bid. *Third World Quarterly* 25:7, 1311–1324.

Swart, K., U. Bob, B. Knott and M. Salie, 2011. A sport and sociocultural legacy beyond 2010: A case study of the Football Foundation of South Africa. *Development Southern Africa* 28:3, 415–428.

UEFA, 2012. *Social Responsibility Report: UEFA EURO 2012 – Creating History Together.* Available at: https://it.uefa.com/MultimediaFiles/Download/uefaorg/General/02/10/87/62/2108762_DOWNLOAD.pdf accessed 11 July 2019.

UN, 2012. *The Future We Want: Rio +20 United Nations Conference on Sustainable Development.* New York: United Nations.

UNOSDP, n.d. *Sport and the Sustainable Development Goals: An Overview Outlining the Contribution of Sport to the SDGs.* Available at: www.un.org/sport/sites/www.un.org.sport/files/ckfiles/files/Sport_for_SDGs_finalversion9.pdf accessed 11 July 2019.

Wackernagel, M. and W. Rees, 1996. *Our Ecological Footprint: Reducing Human Impact on the Earth.* Philadelphia, PA: New Society.

Chapter 3

Golf, the environment, and development

Past, present, and future

Brad Millington and Brian Wilson

Introduction

In 2016, golf returned to the Olympics. It was a much-anticipated event. Golf originally featured as an Olympic event at the 1900 Games in Paris. It was an Olympic event again four years later, but disappeared from the competition schedule after the 1904 Games in St Louis. More than a century later, golf entered the Olympic spotlight again – this time in Rio de Janeiro.

Aside from the excitement of the women's and men's competitions, golf's revival in the context of the 2016 Games sparked interest for at least two reasons. One was that the Olympic Golf Course – built in Rio's Barra da Tijuca region – was touted as a public resource that would catalyse interest in golf in Brazil for generations to come. A second source of excitement was that the new Olympic course was allegedly a feat of environmental engineering. Designed by acclaimed American course architect Gil Hanse, the idea was that, through careful planning, the human-made golf course could make nature *better* (e.g., see Olympic.org, 2015). A post-Games article on the website Olympic.org quoted course director Carlos Favoreto in relation to both of the above-described tropes: 'The great legacy is that any citizen, rich or poor, can come here and get in contact with the sport. This is the true legacy of the golf course, the access to everyone.' The article continued, again quoting Favoreto: '[the course is] the biggest project I've ever seen in terms of protection of fauna and flora environmental and ecosystem recovery' (Olympic.org, 2017).

Golf's return to the Olympics was indeed a much-anticipated event. But there is a far less rosy picture to be painted as well. Less than a year after its opening, news headlines announced the Olympic course's imminent demise (e.g., see Knowlton, 2017; McGonigal, 2017; *The Guardian*, 2017). The theme was roughly the same time and again: the Olympic course, among other venues built for the Rio Games, was the latest inductee into the hall of Olympic 'white elephants', meaning fit-for-mega-event venues that achieved near-instant obsolescence once the Games left town. Industry publications told a similar story. 'Rio Olympic golf course's future looks bleak' was the headline for *Golf Digest* (Beall, 2016). At Golfchannel.com: 'Olympic course not the beacon of hope intended' (Hoggard, 2017). For *Golfweek*, the account was much the same: 'Rio Olympic Golf Course nearly empty, reportedly could "die" in near future' (Casey, 2016).

Indeed, that the Olympic course would be a clear-cut social 'good' was a much-contested claim from the outset. From the development phase onward, the course inspired a groundswell of resistance from the local public, with people rallying together under slogans such as 'ocupa golfe' ('occupy golf') and 'golfe para quem' ('golf for whom?'), and in so doing, effectively rejecting Favoreto's claim that the course was made for everyone to use.

That the course was a clear-cut *environmental* 'good' was and remains a hotly contested notion as well. The Olympic course was not just built in Rio's Barra da Tijuca region. In Watts' (2015) terms, the course 'encroaches' on a nature reserve known as Reserva de Marapendi – a protected environmental site. Critics have argued that claims regarding the course's environmental benefits are dubious, and that the *process* for assessing the ramifications of building and maintaining the course was deeply flawed as well (e.g., see Gordon, 2016; Hodges, 2014; Millington, Darnell & Millington, 2018; Watts, 2015).

Perhaps nature is now striking back. The aforementioned *Golfweek* article ascribes the Olympic course's 'death' to a wildlife 'invasion' made possible by the fact that human golfers were keeping their distance.

This chapter examines the relationship between golf, development, and the environment. We begin with the story of the Rio de Janeiro Olympic course because it is the latest high-profile case of golf being mobilised as an instrument for *sustainable development* – a term that encapsulates economic, social, and environmental sustainability. But we also begin with the Rio case because it shows, in near-perfect terms, the *politics* of sustainable development as they pertain to the sport of golf. On the one hand, what the Rio case demonstrates is the promise that economic, social, and environmental sustainability can be achieved all at once. Said otherwise, the idea is that the golf industry can expand to cover new terrain and can grow its consumer base without delivering undue social and environmental harms – and, indeed, that it can even deliver social and environmental benefits. This viewpoint is commonly evinced by a range of stakeholders, from golf industry representatives to government officials to environmental auditing companies and beyond.

On the other hand, the Rio case shows fault lines in the bedrock of sustainable development. A generous interpretation of events would still give the Olympic Golf Course time to achieve its lofty expectations. Yet, at the very least, it is clear that a golf course cannot be counted on as a failsafe instrument for preserving, let alone improving, either social life or the environment. The headline 'Olympic course not the beacon of hope intended' should inspire consideration of whether *golf in general* is a beacon of hope in the quest for sustainability.

Our argument in this chapter is that the evidence points to a need to *radically* rethink golf's relationship to development and sustainability – for example, by better accounting for local voices on the questions of whether and how golf can be a social and environmental 'good'. To arrive at our main argument, we spend the remainder of this chapter exploring three related questions. First, we ask, why care about the relationship between golf, development, and the environment? What precisely is at stake? The Rio case offers clues, but there is more to be said. We then ask a historical question: How did we get to where we are now? Which is to say, how did we get to the current point whereby golf – a sport that began as a simple and decidedly local game – is perceived as a vehicle for achieving sustainable development in a various contexts around the globe? Finally, we ask, where is golf headed, and what can be done to help golf reach a 'better' place, particularly in terms of its environmental implications?

In developing the analysis below, we draw from our ongoing research into golf and the environment (see Millington & Wilson, 2017, 2016a, 2016b, 2015, 2013; Wilson, 2012; Wilson & Millington, 2013, 2015, 2017). This work has involved a range of research methods and attention to a wealth of golf industry-related resources, including trade publications, government policies, and news media reports. The focus of our research to date has largely been on the American, Canadian, and UK contexts, yet the fact that golf is a globally interconnected game is inescapable, as reflected in the Rio case and in the analysis below.

Golf, the environment, and development – why care?

The first question, then, is why care about the relationship between golf, development, and the environment? The answer is perhaps obvious: that golf can indeed have significant and negative impacts on the environment and human health and wellbeing. Below we address three particular issues: pesticides, water consumption, and the appropriation of land in golf course construction.

Pesticides are perhaps the most common topic of concern in this regard – though, like many issues in this area of study, the 'riskiness' of pesticides is often debated. A pesticide is a substance or mixture of substances 'intended for preventing, destroying, repelling, or mitigating any pest' (US Government Publishing Office, 2013; see also Environmental Protection Agency, 2014). With this in mind, the utility of pesticides on golf courses becomes clear. A lush, green, unblemished course aesthetic has become the gold standard in golf (a point we reflect on in further detail below). Pesticides offer efficient means of eliminating the insects (insecticides), fungus (fungicides), and weeds (herbicides) that threaten attempts to attain this standard.

In Canada and the US, it is now widely accepted that, *in the past*, uses of pesticides on golf courses posed far too great a risk to the environment and human health than what was merited given the upside of pristine course conditions. Indeed, this point is sometimes articulated by people working *within* the golf industry, and not just by external critics. For example, as said by prominent course architect Mike Hurdzan in a feature in *Golf Digest*:

> Back in the mid-'50s we were using cadmium, lead, arsenic, mercury; we were using all these heavy metals. We were using farm-grade fertilizers. Well, those things are gone. We didn't know any better back then. Science has showed us a better way to do things.
>
> (Barton, 2008)

In this same article, Hurdzan draws a distinction between pesticides-as-poison and pesticides-as-medicine. His assertion that science has shown us a 'better way' is an indication that we have moved from the former scenario to the latter.

But this is where controversy arises. There is a lingering concern that we are repeating the mistakes of the past, even if it is true that the golf industry has done away with the heavy metals of the 1950s and other chemicals that are now widely understood to be too risky for golf. In 2004, for example, Knopper and Lean (2004) concluded, based on their review of relevant research, that 'There appears to be convincing evidence to support the claim that under certain circumstances [pesticides commonly used on golf courses] have been associated with cancer' (p. 276). They go on to acknowledge that the associations in question are usually weak, but that, even so, a weak association is different from no association at all: 'studies presenting these results should not simply be … misinterpreted as meaning that exposure to the compound in question is not related to any health concern' (p. 276).

More recently, health and environmental organisations such as the Canadian Cancer Society and the David Suzuki Foundation have come out in strong terms against the sustained use of chemicals on golf courses in Canada. The backdrop for this is that many Canadian provinces have banned cosmetic pesticide usage, but have given golf a legislative exemption. For the Canadian Cancer Society, 'The research shows that pesticide exposure in general (as well as exposure to specific pesticides) is linked to several types of cancer such as non-Hodgkin lymphoma, multiple myeloma, and prostate, kidney and lung cancers, among others' (Canadian Cancer Society, 2018a). With this general perspective in mind, it makes sense that the Canadian Cancer Society would specifically recommend – contra golf's exemption from provincial legislation – that pesticides be phased out on golf courses and other sports facilities (Canadian Cancer Society, 2018b).

These concerns pertain strictly to humans. There are reasons for trepidation over the impacts of pesticides on wider ecosystems as well. Neonicotinoids offer one case in point. Neonicotinoids are pesticides that act on the neural system of invertebrates; they have grown increasingly prominent since their commercialisation in the early 1990s. Recently, however, researchers and health and environmental organisations have sounded alarm bells over the consequences of the widespread adoption of these chemicals, the most disconcerting of which involves their apparent impacts on bees (see Carrington, 2018; European Food Safety Authority, 2018; Task Force on Systemic Pesticides, 2018). The city of Montreal's 2015 ban on neonicotinoids, *including on golf courses*, is perhaps a sign of things to come (CBC News, 2015).

Course irrigation is a second issue when it comes to the environmental implications of golf. There is no straightforward response to the question of how much water golf courses require; the answer depends, for example, on where the course is located (e.g., due to annual rainfall) and the desired aesthetic among ownership and staff. The type of water used is a consideration too – for example, the golf industry has highlighted how recycled water is sometimes used in golf course maintenance (Golf Course Superintendents Association of America [GCSAA], 2009b).

But the point, again, is that golf is *potentially* burdensome in this regard – a fact that sometimes receives press attention when drought conditions strike. For instance, amidst a recent drought in California, golf courses were compelled to make significant cuts in their water usage (Ca.gov, 2015). In this context, one media report put the figure for annual water use on US golf courses at over 100 billion gallons per year – more than offices, schools, restaurants, and other non-agricultural industries. By the industry's own figures, from 2003 to 2005 US golf courses used more than 2 billion gallons of water per day for course irrigation (GCSAA, 2009b).

This is to say nothing of what the global expansion of the game means for resource consumption, especially as golf moves to arid climates. In 2010, *The Guardian* reported that golfer Tiger Woods's planned resort course in Dubai required a million gallons of water a month to keep vegetation alive while the course was in construction (Donegan, 2010).

Finally, there is the matter of land appropriation in course development – and what this does to people, wildlife, and ecosystems such as ponds and swamps. In a 2007 survey, the total size of an average US golf course was found to be 150 acres, roughly two-thirds of which is maintained turfgrass (GCSAA, 2009a). Acreage can, of course, vary depending on context, but to state the obvious, golf is inherently expansive, and golf course construction potentially intrudes on sensitive terrain.

Trump International Golf Links Scotland (TIGLS) is an exemplary case in this regard (Millington & Wilson, 2017). In the late 2000s, the proposal to build the TIGLS course earned approval after a contentious government review process. At the heart of the matter was the issue of sustainable development. On the side of environmental sustainability, there was great concern that the course would stabilise the Foveran Links sand dunes – officially classified as a Site of Special Scientific Interest – given its proposed placement on the Scottish coastline near Aberdeen. But the Scottish Government was compelled by the economic and social cases for going ahead with the course, and so, in the end, the environmental arm of sustainable development lost out. A protest movement called Tripping Up Trump emerged to prevent the displacement of local residents (who accused the Trump organisation of bullying tactics) and to prevent the development of the course altogether. While the movement was unsuccessful in grinding development to a halt, in more recent years the decision to approve the course has inspired regret, including from former First Minister of Scotland, Alex Salmond – who was a fervent supporter of the development at an earlier point in time (Campbell, 2017).

It is important to note here that TIGLS is but one high-profile case. A resort course that was planned in the coastal village of Hacienda Looc in the Philippines is another. The plan in this case was reportedly to build four championship courses. The problem, per journalist Todd Pitock's (2009) account, was as follows:

> No one had bothered to ask the 7,000 farmers, fishermen and villagers who faced eviction from ancestral lands in what the project's critics say amounted to a nefarious land grab orchestrated by corrupt officials and unscrupulous developers.
>
> (p. 91)

Again, a protest movement – in this case called Break Free – formed in response to the proposed development. Pitock's harrowing account enumerates the conscription of military, para-military, and police forces in the face of protests and the death of three of Break Free's leaders by gunfire. To be sure, the Hacienda Looc case, in Pitock's recounting, is extreme, but in general, concerns about the social and environmental dimensions of sustainability are suffused throughout the critical literature on golf (e.g., see Briassoulis, 2010; Neo, 2010; Perkins, Mincyte & Cole, 2010; Stoddart, 1990; Stolle-McAllister, 2004; Wheeler & Nauright, 2006).

Of course, to the question 'why care about the relationship between golf, development, and the environment?', the golf industry would surely mount a different response to that outlined above. Rather than highlighting the potential negative implications of golf course development and maintenance for the environment and human health and wellbeing, we would expect golf industry representatives to emphasise the *positive contributions* golf might make in these areas. Indeed, a common refrain in industry documents is that golf courses preserve wildlife and green spaces alike, that they can 'rehabilitate' unused terrain (e.g., old industrial sites being remade into golf courses), and that golf has value as a healthy leisure pursuit (e.g., see http://wearegolf .org/). This messaging comes in addition to the aforementioned ideas that pesticides are now less risky than in the past and are ultimately beneficial (pesticides-as-a-medicine) and that maintenance staff are becoming savvier in their use of resources like water.

Said another way, a viewpoint emanating from the golf industry – what we would argue is the dominant corporate viewpoint – is that golf is a force for good in the quest for sustainable development. The Golf Course Superintendents Association of America lays this out in explicit terms in describing the ideal relationship between golf and the environment: 'Sustainability focuses on the "triple bottom line" – people, planet, and profit – to ensure businesses are successful' (GCSAA, 2018).

It is true that there can be valid points on both sides of the argument – that golf can be environmentally detrimental in some ways or in some circumstances, but beneficial in others. But the overriding point from a critical perspective is that industry claims around environmental leadership should not be taken at face value. Sustaining the planet and sustaining an economic system are not the same thing, and should not be put on equal footing as priorities – particularly when the economic system in question is one that is associated with the rise and perpetuation of pressing environmental problems (see Foster, Clark & York, 2011; Magdoff & Foster, 2010). And, of course, there is evidence that environmental and economic concerns are not even on equal footing – considering golf's environmental demands, the sustained use of particular chemicals, and particular cases whereby 'the planet' has seemingly been a secondary concern at best.

How did we get here?

So, how did we arrive at a place where the golf industry, in the face of ongoing criticism, has positioned itself as a leading force on environmental issues? The answer is one that maps onto a broader story of corporate environmentalism (see Hoffman, 2001).

What our research has shown is that the dominant perspective on development in the golf industry has not always been development-as-sustainable-development (Millington & Wilson, 2016b). In past eras, an emphasis lay on development-as-modernisation, with modernisation meaning the adoption of technology in the interest of *controlling* the environment. Our evidence for this claim comes in the form of industry trade publications, meaning materials made by and for golf industry representatives. For example, in a publication called *The Golf Course* from the early 1900s, we found proclamations that 'modern' golf courses can and should surpass their predecessors in America – and even those in golf's British homeland. Take, for instance, the viewpoint of famed course architect A.W. Tillinghast (1916), expressed in a repeating feature in *The Golf Course* called 'Modern Golf Chats':

> It is not necessary to attempt a description of those early American courses, with their featureless greens, mathematically correct and symmetrical bunkers and the ridiculous little bandbox teeing grounds. They are of the past, but they served their purpose. The golf courses which we Americans are constructing to-day are very different, and so carefully are they built, after a thoughtful preparation of plans, that some of our productions are not surpassed even in the old home of golf.
>
> (p. 6)

It was not long before pesticides were added to the equation. When combined with technologies for radically manipulating the land – earth-moving machinery and explosives, for example – people were, by the inter-war years, empowered to manipulate the environment in a way unlike ever before.

We have seen already how golf architect Mike Hurdzan has lamented past uses of chemicals in the alleged pesticides-as-poison era. The chemical DDT (dichlorodiphenyltrichloroethane) is probably the best case in point in this regard. Designed in the World War Two years for the purpose of insect control, DDT is highly potent. What it offered the golf industry was the opportunity to press further than ever in the quest to eradicate unwanted pests. In trade publications from the post-war years, we find stories of indiscriminate DDT spraying by various means, including, in one case, via helicopter (Anon., 1967; see Millington & Wilson, 2013).

But chemicals of this kind also created controversy. The development-as-modernisation ethos is perhaps best articulated by then-GCSAA president Richard C. Blake in a trade publication 'President's Message' from 1971:

> The truth of the matter is that civilization did not begin until man learned to use fire and other tools to modify his environment. In other words, the fate of the human race and the wildlife that has shared in its rise rests on man's ability to anticipate, modify, and control environmental *changes*.
>
> (Blake, 1971, p. 7, emphasis in original)

The value of nature is thus reduced to its value to humankind.

At roughly this same moment, however, the environmental movement was gathering steam. In her famous book, *Silent Spring*, which helped catalyse interest in the environment, Rachel Carson (2002 [1962]) noted how casual pesticide use had become:

> If we are troubled by mosquitoes, chiggers, or other insect pests on our persons we have a choice of innumerable lotions, creams, and sprays for application to clothing or skin ... To make certain that we shall at all times be prepared to repel insects, an exclusive New York store advertises a pocket-sized insecticide dispenser, suitable for the purse of for beach, golf, or fishing gear.
>
> (p. 175)

DDT in particular was an object of Carson's and other environmentalists' attention.

While it is impossible to know the exact extent to which golf industry representatives were concerned with the rising tide of the environmental movement – and recognising that the influence of this movement was not distributed in perfect evenness around the globe – it seems that the status quo of development-as-modernisation was untenable. Indeed, in North America at least, by the 1980s the golf industry was moving in a new direction: towards the development-as-sustainable-development paradigm that is the norm today. Pesticides again offer a case in point. In one sense, there was a drive to move to less risky pesticides – to make the transition from 'poison' to 'medicine', as we have already seen. At the same time, there was a push, at least in rhetoric, to use pesticides *less often*. In the 1980s, a system called Integrated Pest Management (IPM) was adopted from agriculture as a way of moving this agenda along. In short, IPM involves using pesticides as a 'last resort' once other, non-chemical approaches are tried (e.g., dealing with pests by introducing predator species to an ecosystem).

The golf industry was moving from the back foot to the front: the message was and remains that golf industry representatives are environmental *leaders*; they are themselves environmentalists. Again, we cannot say for certain that this was devised as a political strategy in the main, but it clearly had political ramifications.

As we have discussed elsewhere (Millington & Wilson, 2016a), from the 1980s onwards, many governments were *looking* for environmental leaders in industry. A key legislative strategy that emerged around this time – one called 'environmental managerialism' by the sociologist John Hannigan (2006) – is for governments to express a commitment to protecting the environment while at the same time preserving a focus on economic growth. The concept of sustainable development, with its emphasis on the 'triple bottom line', is perfectly suited to this environmental man-agerialist approach. By professing a commitment to environmental leader-ship, businesses could appease governments that might, under other circumstances, impose highly restrictive environmental legislation; govern-ments were not ignoring their responsibility for environmental protection, but rather passing the burden to industry experts.

We have argued that an archetypal case of environmental managerialism (from a government perspective) and sustainable development (from a corporate perspective) fitting together like lock and key is the aforemen-tioned banning of pesticides through provincial legislation in Canada. We have looked most in-depth at the province of Ontario, where the 2009 ban on cosmetic uses of pesticides (e.g., on household lawns) was heralded as landmark legislation by environmental organisations such as the David Suzuki Foundation. Yet the law was also critiqued for exempting golf. Part of what helped the golf industry's case for side-stepping this legislation were the environmental 'best practices' such as IPM that, by the 2000s, had become commonplace (Millington & Wilson, 2016a). In other words, this is an instance of environmental leadership having political benefits.

So how did we get to the current point where golf is perceived as a vehicle for achieving sustainable development in various contexts around the globe? The development-as-modernisation approach, whereby 'civilisation' and environmental control are conjoined, was untenable in the face of the environmental movement. But the upshot was not to aban-don modernisation; it was to embrace it in a new form. This is *ecological* modernisation. The idea was and is that technology and innovation – for example, better and more strategically used pesticides, water-saving tech-niques, precision spraying, and careful construction through knowledge of the land – could pave a pathway towards sustaining the environment while still allowing the industry to grow.

We return here to the 2016 Olympics, and to the narrative that a human-made construction – despite its seeming environmental demands and risks – can make both nature and social life better. One can imagine that, had golf returned to the Olympics 50 years sooner, the rationale for building a new course for the Games would be very different, and would not contain such a strong focus on the 'triple bottom line'. That the Olympic course was promoted as socially and environmentally beneficial is in keeping with wider changes in golf.

Conclusion: where is golf headed?

Protecting the environment and promoting human health and wellbeing require commitment from everyone – from the public sector to industry to the population at large. When it comes to the environment, golf in many ways is in a better place than it was in the early post-war years, meaning in the heyday of the development-as-modernisation era. This should not be easily dismissed.

At the same time, we would argue that much more could be done in golf from an environmental perspective. The problems associated with golf course construction and maintenance have not gone away, even with the golf industry's adoption of corporate responsibility. For example, the pesticides-as-poison/pesticides-as-medicine binary is a flawed metaphor. It ignores how chemicals once thought to be safe (or at least to pose an acceptable level of risk) have, with the passage of time, inspired a great deal of concern (see Millington & Wilson, 2013). It also ignores that medicine itself can become poison, depending on the dosage.

We can also add here that there remain strong incentives in place for people such as golf superintendents to strive towards a 'perfect' golf course aesthetic. The idealised picture of the lush, green, unblemished golf course is a remnant of the development-as-modernisation era, when pests had little hope against powerful pesticides that were sometimes applied indiscriminately. This picture of the highly manicured golf course is a hard one to undo.

What is more, embracing sustainable development has helped the golf industry achieve a place of relative autonomy. The industry is certainly not free from regulatory constraints (and the regulatory picture, of course, differs across the globe). But industry representatives are often situated in contexts where government officials are keen for businesses to 'lead the way' on environmental issues. What this has helped engender are cases where the three components of the 'triple bottom line' are unequally weighted, with economic growth receiving priority treatment. In a different set of circumstances, environmentalists' warning that a golf course would (for example) irreparably damage a protected Site of Special Scientific Interest would be all that is needed by way of a sustainability assessment.

Where is golf headed? We would argue that where it *should* be headed is a place where the environmental dimension of sustainable development is given much greater weight. In the most optimistic assessment, 'darker green' versions of sports like golf might inspire changes in the wider society. Arthur Mol (2010) – a key intellectual figure behind the concept of ecological modernisation – has argued that sustainability is a 'global attractor', meaning it is a concept that 'redirects institutions, practices, structures, norms and ideologies globally' (p. 511) towards improved environmental outcomes. Mol (2010) further contends that mega-events are telling exemplars in this regard; he uses the 2008 Beijing Olympics as a case in point. China, Mol argues, aspired to host the Olympics to legitimise its position in the international community; the country's leadership needed to show concern for the environment to achieve this end. As Mol (2010) writes, 'The global and domestic interests and legitimacy of the [Chinese Communist Party] and the state are to a major extent fortified by a better environmental record and more environmental transparency' (p. 524).

In fairness, Mol (2010) recognises that the Beijing Games have been criticised as well, including for their environmental implications. But his account is ultimately optimistic about the potential of sustainability as a concept, and the potential for sport to drive sustainability forward.

As should be clear by now, we do not share Mol's optimism. It is worth noting that there is evidence that measures taken around the time of the 2008 Games to (for example) improve air quality did not have staying power (e.g., see Chen et al., 2013), and that subsequent Olympics have been criticized for being environmentally damaging. Sustainability is indeed attractive, but this is precisely because of its fungible nature – which also makes its environmental component expendable, or at least diminishable.

What we share with Mol, however, is hope that consideration of the environment can indeed serve as a propulsive force for 'institutions, practices, structures, norms and ideologies', and that sport can play an important role. It is just that a 'darker green' conception of the environment is needed. What exactly 'dark green' sport comprises remains an open question, and thus a priority for researchers. Research no doubt must retain a focus on critique, and must be broad in its scope. We acknowledge that our view that the golf industry has adopted a position of corporate responsibility in recent years is based on our assessment of English-language materials, and pertains in large part to the west (and especially Canada and the United States). The rhetorical question emanating from Rio de Janeiro – Golf for whom? – should compel researchers to continue considering questions of power and inequality in different contexts around the globe.

But if golf is to move to a 'better place', and if sport in general is to move in a similar direction, researchers must also help illuminate cases that go beyond the corporate environmentalist norm. As noted, there are examples of pesticide-free golf courses and of courses that are built on once-industrial sites. But these 'dark green' alternatives have received very little attention in the literature or in media – and so we know little about their scalability and, in turn, about their ultimate potential. The 'coping' mechanisms enacted in extreme environmental circumstances (e.g., drought) might also provide a template for a more sustainable status quo in the years ahead.

Mol (2010) is not alone in advocating for the power of sport. The United Nations, for example, positions sport as an important enabler in achieving its 2030 Agenda for Sustainable Development. We might imagine, down the line, a sport mega-event where the environmental ramifications of hosting the event are reasons, *by themselves*, for not hosting at all, or a mega-event where no new facilities are built (we would do well to remember here that Rio de Janeiro already had two golf courses before building a new one for the games (Gordon, 2016)). Or we might imagine a time when the norm is to give community members an authentic voice in whether a golf course should be built – and if the verdict is 'yes', that these same community members would be given a determining say in the environmental principles by which the course should abide.

In all, a darker green version of sustainability would mean privileging the environment based on its inherent value – and not its value relative to other forms of sustainability. Darker green sport might inspire thinking and action on darker green outcomes in general.

References

Anon. (1967). Golf course chemical warfare takes to the air. *Golf Superintendent*, July, 8–9.

Barton, J. (2008). The golf-course architect: Mike Hurdzan. How green is golf? Retrieved 14 April 2018 from www.golfdigest.com/story/environment_hurdzan.

Beall, J. (2016). Rio Olympic golf course's future looks bleak. Retrieved 14 April 2018 from www.golfdigest.com/story/rio-olympic-golf-courses-future-looks-bleak.

Blake, R.C. (1971). Reason over emotion. *Golf Superintendent*, March, 7.

Briassoulis, H. (2010). 'Sorry golfers, this is not your spot!': Exploring public opposition to golf development. *Journal of Sport & Social Issues*, 34(3), 288–311.

Ca.gov. (2015). Governor Brown directs first ever statewide mandatory water reductions. Retrieved 14 April 2018 from www.gov.ca.gov/2015/04/01/news18913/.

Campbell, G. (2017). Alex Salmond: Trump 'broke Scots investment promises'. Retrieved 14 April 2018 from www.bbc.com/news/uk-scotland-41918224.

Canadian Cancer Society. (2018a). Pesticides and cancer. Retrieved 14 April 2018 from www.cancer.ca/en/prevention-and-screening/reduce-cancer-risk/make-informed-decisions/be-safe-at-work/pesticides-and-cancer/?region=on.

Canadian Cancer Society. (2018b). The Canadian Cancer Society's perspective on pesticides. Retrieved 14 April 2018 from www.cancer.ca/en/prevention-and-screening/reduce-cancer-risk/make-informed-decisions/be-safe-at-work/the-canad ian-cancer-societys-perspective-on-pesticides/?region=on.

Carrington, D. (2018). Total ban on bee-harming pesticides likely after major new EU analysis. Retrieved 14 April 2018 from www.theguardian.com/environment/2018/feb/28/total-ban-on-bee-harming-pesticides-likely-after-major-new-eu-analysis.

Carson, R. (2002 [1962]). *Silent spring*. New York: Houghton Mifflin.

Casey, K. (2016). Rio Olympic Golf Course nearly empty, reportedly could 'die' in near future. Retrieved 14 April 2018 from http://golfweek.com/2016/11/25/rio-olympic-2016-golf-course-empty-could-die-brazil/.

CBC News. (2015). Montreal bans neonicotinoid pesticide to help save the bees. Retrieved 14 April 2018 from www.cbc.ca/news/canada/montreal/montreal-bans-neonicotinoid-pesticide-to-help-save-the-bees-1.3360458.

Chen, Y., Jin, G.Z., Kumar, N., & Shi, G. (2013). The promise of Beijing: Evaluating the impact of the 2008 Olympic Games on air quality. *Journal of Environmental Economics and Management*, 66(3), 424–443.

Donegan, L. (2010). Tiger Woods's Dubai dream evaporates in the desert. Retrieved 14 April 2018 from www.theguardian.com/sport/2010/nov/28/tiger-woods-golf-course-dubai.

Environmental Protection Agency. (2014). About pesticides. Retrieved 8 April 2015 from www.epa.gov/pesticides/about/.

European Food Safety Authority. (2018). Neonicotinoids: Risks to bees confirmed. Retrieved 14 April 2018 from www.efsa.europa.eu/en/press/news/180228.

Foster, J.B., Clark, B., & York, R. (2011). *The ecological rift: Capitalism's war on the earth*. New York: New York University Press.

GCSAA. (2009a). Golf course environmental profile. Volume I. Property profile and environmental stewardship of golf courses. Retrieved 14 April 2018 from www.gcsaa.org/environment/golf-course-environmental-profile.

GCSAA. (2009b). Golf course environmental profile. Volume II. Water use and conservation practices on U.S. golf courses. Retrieved 14 April 2018 from www.gcsaa.org/environment/golf-course-environmental-profile.

GCSAA. (2018). Sustainability. Retrieved 14 April 2018 from www.gcsaa.org/environment/sustainability.

Gordon, A. (2016). Rio didn't need an Olympic Golf course, but they built one anyway. Retrieved 14 April 2018 from https://sports.vice.com/en_ca/article/wnmjzq/rio-didnt-need-an-olympic-golf-course-but-they-built-one-anyway.

The Guardian. (2017). Rio Olympic venues already falling into a state of disrepair. Retrieved 16 April 2018 from www.theguardian.com/sport/2017/feb/10/rio-olympic-venues-already-falling-into-a-state-of-disrepair.

Hannigan, J. (2006). *Environmental sociology*, 2nd edition. New York: Routledge.

Hodges, E. (2014). The social and environmental costs of Rio's Olympic golf course. Retrieved 14 April 2018 from www.rioonwatch.org/?p=17283.

Hoffman, A.J. (2001). *From heresy to dogma: An institutional history of corporate environmentalism*. Palo Alto, CA: Stanford University Press.

Hoggard, R. (2017). Olympic course not the beacon of hope intended. Retrieved 14 April 2018 from www.golfchannel.com/article/rex-hoggard/olympic-course-not-beacon-hope-intended/.

Knopper, L.D. & Lean, D.R. (2004). Carcinogenic and genotoxic potential of turf pesticides commonly used on golf courses. *Journal of Toxicology and Environmental Health, Part B*, 7(4), 267–279.

Knowlton, E. (2017). Here is what the abandoned venues of the Rio Olympics look like just 6 months after the games. Retrieved 18 April 2018 from www.businessinsider.com/rio-olympic-venues-are-abandoned-just-6-months-after-games-2017-2.

Magdoff, F. & Foster, J.B. (2010). What every environmentalist needs to know about capitalism. *Monthly Review*, 61(10), 1.

McGonigal, C. (2017). It's been just 7 months since the Rio Olympics, and this is what the venues look like now. Retrieved 18 April 2018 from www.huffingtonpost.com/entry/rio-olympics-abandoned-photos_us_58d13655e4b00705db52b97d.

Millington, B. & Wilson, B. (2013). Super intentions: Golf course management and the evolution of environmental responsibility. *Sociological Quarterly*, 45(3), 450–475.

Millington, B. & Wilson, B. (2015). Golf and the environmental politics of modernization. *Geoforum*, 66, 37–40.

Millington, B. & Wilson, B. (2016a). An unexceptional exception: Golf, pesticides, and environmental regulation in Canada. *International Review for the Sociology of Sport*, 51(4), 446–467.

Millington, B. & Wilson, B. (2016b). *The greening of golf: Sport, globalization and the environment*. Manchester, UK: Manchester University Press.

Millington, B. & Wilson, B. (2017). Contested terrain and terrain that contests: Donald Trump, golf's environmental politics, and a challenge to anthropocentrism in physical cultural studies. *International Review for the Sociology of Sport, 52* (8), 910–923.

Millington, R., Darnell, S., & Millington, B. (2018). Ecological modernization and the Olympics: The case of golf and Rio's 'green' Games. *Sociology of Sport Journal, 35*(1), 8–16.

Mol, A.P. (2010). Sustainability as global attractor: The greening of the 2008 Beijing Olympics. *Global Networks, 10*(4), 510–528.

Neo, H. (2010). Unravelling the post-politics of golf course provision in Singapore. *Sport and Social Issues, 34*(3), 272–287.

Olympic.org. (2015). Rio 2016 golf course unveiled. Retrieved 14 April 2018 from www.olympic.org/news/rio-2016-golf-course-unveiled.

Olympic.org. (2017). Bright future predicted for the golf course of the Olympic Games Rio 2016 one year later. Retrieved 14 April 2018 from www.olympic.org /news/bright-future-predicted-for-the-golf-course-of-the-olympic-games-rio-2016-one-year-later.

Perkins, C., Mincyte, D., & Cole, C.L. (2010). Special issue: Making the critical links and the links critical in golf studies. *Journal of Sport & Social Issues, 34*(3), 267–375.

Pitock, T. (2009). Turf wars. *Australia Golf Digest*, 90–94.

Stoddart, B. (1990). Wide world of golf: A research note on the interdependence of sport, culture, and economy. *Sociology of Sport Journal, 7*(4), 378–388.

Stolle-McAllister, J. (2004). Contingent hybridity: The cultural politics of Tepoztlán's anti-golf movement. *Identities: Global Studies in Culture and Power, 11*, 195–213.

Task Force on Systemic Pesticides. (2018). Systemic pesticides. Retrieved 14 April 2018 from www.tfsp.info/systemic-pesticides/.

Tillinghast, A.W. (1916). Modern golf chats. *The Golf Course*, January, 1–7.

US Government Publishing Office. (2013). United States Code. Title 7 – Agriculture. Retrieved 14 April 2018 from www.govinfo.gov/app/collection/uscode/2013/.

Watts, J. (2015). Rio 2016: 'The Olympics has destroyed my home'. Retrieved 14 April 2018 from www.theguardian.com/world/2015/jul/19/2016-olympics-rio-de-janeiro-brazil-destruction.

Wheeler, K. & Nauright, J. (2006). A global perspective on the environmental impact of golf. *Sport in Society, 9*(3), 427–443.

Wilson, B. (2012). Growth and nature: Reflections on sport, carbon neutrality, and ecological modernization. In D. Andrews & M. Silk (Eds.), *Sport and neoliberalism: Politics, consumption, and culture* (pp. 90–108). Philadelphia, PA: Temple University Press.

Wilson, B. & Millington, B. (2013). Sport, ecological modernization, and the environment. In D. Andrews & B. Carrington (Eds.), *A companion to sport* (pp. 129–142). Malden, MA: Blackwell.

Wilson, B. & Millington, B. (2015). Sport and environmentalism. In R. Giulianotti (Ed.), *Routledge handbook of the sociology of sport* (pp. 366–376). New York: Routledge.

Wilson, B. & Millington, B. (2017). Sport, the environment, and physical cultural studies. In M. Silk, D.L. Andrews, & H. Thorpe (Eds.), *Routledge handbook of physical cultural studies* (pp. 333–343). New York: Routledge.

Chapter 4

Approaching sport and the environment through immersive education technologies

Kyle Bunds

Introduction

In June 2017 at the *Sport and Sustainable Development: Setting a Research Agenda* symposium in Toronto that precipitated this book, discussions centered on how we might best understand the intersecting roles and impacts of sport, the environment, and development. Several overarching research questions arose, including: How should we define sport and the environment within development? Does sport offer something unique to conventional approaches to development? And what should be the goal of this research agenda going forward? Such questions have guided my various research projects for several years.

With these questions in mind, this chapter focuses on some of the ways in which education may serve to bridge sport, environment, and development[1] in productive ways. Specifically, I argue that utilizing sport in support of environmental philanthropy and education, and particularly engaging people through immersive education technologies within a sport setting, can help draw attention to some of the most serious environmental issues facing our times, while capturing a broad audience of sport program participants. The chapter therefore serves as a proposal of sorts, one focused on how sport might fit into the challenge and practice of promoting environmentally sustainable development.

To that end, I offer reflections from more than five years of research examining the activities of development agencies that use sport as a way to raise funds in order to build water systems in developing countries (Bunds, 2017). This research has found that sport can raise funds and encourage involvement in charitable efforts on the one hand, but on the other, can also reinforce hierarchies and unequal power relations between the global North and global South (also see Darnell, 2007; Hayhurst, 2011).

These reflections lead to and illustrate three main ideas that have emerged from this research: (1) sport can be used to increase people's interests in social causes, including environmental issues, a process that to date has been relatively effective, if sometimes problematic; (2) marketing a story to consumers is a way that organizations can create loyal donors, but also create citizens who take an ongoing interest in environmental issues; and (3) immersive educational technologies can be particularly beneficial in educating individuals about the environment, especially in a sport setting or by utilizing the popularity of sport itself. Before exploring these ideas in more detail, I offer a brief review of literature on sport, development, and social change, as well as a short discussion of environmental education and informal learning spaces.

Sport, development, and social change

Overall, research on the role of sport in development and social change is mixed. Some sports scholars argue that sport has the ability to positively influence individuals (Jones et al., 2018a), with most of these arguments made in the youth development literature (Jones et al., 2018b). Specifically, scholars contend that sport can lead to positive outcomes among youth participants in relation to social cohesion, community building, and civic participation (Spaaij & Jeane, 2013). Skinner, Zakus, & Cowell (2008) also found that sport can be an important avenue for building community. In particular, they found that the promotion of sport among community members may help build continuity and bring people closer together.

Two specific aspects of the relationship between sport and social change are particularly relevant to this chapter. The first is that sport may inspire people to act in the service of social issues and causes. Kaufman and Wolff (2010) have suggested that sport offers a space and means for social activism, and that such activism can be powerful in its impact and effects. Kaufman and Wolff's (2010) research in turn substantiates the work of Wright and Coté (2003), who also found that sport can be an important part of building and developing individuals who are socially aware. Particularly, they found that sport helped people to develop leadership skills, which can be valuable individually and socially. Second, Spaaij and Jeane (2013) have argued that sport can serve as a tool for educating people about social issues and processes of social change. To do so, however, they argue for a move beyond traditional ways of teaching, such as didactic interaction, and toward new styles of engagement. In sum, sport interventions would seem to hold potential for educating individuals about important issues – such as the environment – if done creatively and correctly (Bunds, 2017) and with the appropriate organizational capacity (Edwards, 2015).

At the same time, critical assessments of sport, development, and positive social change have suggested that sport's development potential is often presumed or taken for granted rather than proven (Coalter, 2007).

Spaaij, Farquharson, and Marjoribanks (2015), for example, found that sport can exist as a site for inclusion, but can also lead to exclusion if not properly managed, and may even reinforce hierarchies of gender, race, and national identity. Such critical perspectives are not new in the socio-logical analysis of sport, but confirm long-standing views of sport as a site of power and the construction and maintenance of ideology. As Sage (1990) explained:

> Sport is considered to be an important site upon which dominant ideology is constructed and maintained because sport's institutional and ideological features have evolved in a way that corresponds with, and helps to reproduce, the conditions upon which dominant interests are based.
>
> (p. 26)

In the context of this chapter, what these insights mean is that the "how" of environmental education through sport is as important as the "what." With this in mind, I turn next to some key issues from the field of environmental education.

Environmental education

The question of how to frame social phenomena like sport in meeting environmental and development goals is not restricted to the sociology of sport and sport management; indeed, environmental education scholars have also struggled with what is reasonable and possible. In the 1970s, when environmental education first became part of university curricula, it was used mainly as a tool to contextualize scientific phenomena (Tilbury, 1993). The fledgling field of environmental education remained as such well into the 1990s, when the idea of Environmental Education for Sustainability (EEFS) was introduced as a way to consider "how people interact with their total environment and with addressing environmental problems holistically through curriculum" (Tilbury, 1995, p. 199). This new idea sought to encapsulate the socio-historical elements of the environment, and also incorporated the idea of sustainability which had gained traction in the 1980s after the World Conservation Strategy defined the term and the 1987 Brundtland Report reinforced its importance. Other scholars, like Sauve (1996), eventually built on Tilbury's (1995) work to suggest that EEFS should be a part of the broader scope of educating responsible citizens who could positively contribute to sustainability.

This move toward a more holistic understanding of environmentalism informed, and was informed by, EEFS and was also part of the growing recognition of the need to include sustainability in approaches to development. Sustainable development became a key feature of the World

Conservation Strategy of 1980 proposed by the International Union for Conservation of Nature, United Nations Environment Programme, and World Wide Fund for Nature (IUCN/UNEP/WWF, 1980), and the connections between poverty, development, and the environment were considered in relation to one another in more explicit terms (Tilbury, 1995). As with the shift toward EEFS around the same time, global environmental conferences in the 1980s and into the early 1990s foregrounded education as key to solving issues of poverty, development, and the environment. As Tilbury argues, the 1991 World Conservation Strategy (IUCN/UNEP/WWF, 1991) and the UN *Conference on Environment and Development* of 1992 firmly entrenched environmental education as a necessary component for sustainable development.

With that said, and despite the well-established links between environment, sustainability, and development, there remains significant resistance within some educational settings regarding the impacts of the environment on sustainable development, and even the anthropogenic causes of climate change. For example, Stevenson, Peterson, and Bradshaw (2016) found that 92% of middle school science teachers in North Carolina believed climate change is real, but only 12% believed it is anthropogenic, while the majority of students believed in anthropogenic causes. In general, this finding is in lock step with a consensus of popular press polls showing that 18–34-year-olds believe more in climate change than any other adult age group and are by far the least skeptical (Saad, 2017). These results show that resistance to ideas about climate change differs by age group, further illustrating the importance of education.

Crucially, then, it is important to note that researchers have found informal, activity-based environmental education, especially through place-based learning opportunities, to be an effective means for enhancing what students learn in the classroom (Woodhouse & Knapp, 2000). Thus, there is a need to understand informal spaces, such as those provided by sport, in relationship to environmental education. In their 2017 STEM Advancing Informal Science Learning call for proposals, the National Science Foundation (NSF) of the United States described informal learning spaces in the following terms:

> Almost any environment can support informal science learning, such as a home, a museum, a library, a street, or a virtual or augmented reality game. Information networks, mobile media, and social networks transform educational possibilities and create opportunities for seamless learning environments.
>
> (NSF, 2017)

In sum, given that sport can support positive development, especially among youth, and that it may serve as an informal setting for education, particularly focused on the environment, I argue that sport can be utilized as a means through which to engage people, particularly youth, in discussions, learning, and action regarding serious sustainability issues. To that end, next I offer reflections from my research into international development agencies that focus on water issues. I then combine these reflections with the literature discussed here to propose that immersive education technologies might be utilized to position sport for environmental education in support of sustainable development.

Water charities, fundraising, and logics of individualism

Throughout my time working with different water development agencies, like London Water Charity,[2] one consistent feature was the importance for these organizations to communicate strong messages that connect with people (Bunds, 2017). London Water Charity is a non-profit organization that operates out of London, England and provides water systems to those in need in Malawi and Zimbabwe. London Water Charity has succeeded in raising money for different global water projects, due in large part to their effective message framing, and by partnering with an organization called Charity: Water to produce high-quality videos, often narrated by celebrities such as Kristen Bell.

The Charity: Water website (https://uk.charitywater.org/) is full of suggestions for how "you" can get involved with the move toward fixing the world's water crisis in developing countries (including, for example, purchasing $40 t-shirts with a graphic on the front, which are marketed through Saks Fifth Avenue's association with the charity). Specifically, there are four main areas where "you" can help, according to Charity: Water. "You" can donate money; purchase clothing, playing cards, iPad covers, and many other material goods, volunteer with the organization, and/or fundraise through biking, running, swimming, kayaking, or other physical activities. Notably for the field of sport and physical culture, it is the last of these options that Charity: Water advocates for the most strongly because it focuses the most on "you" as the active benefactor, and even savior.

Indeed, according to its website, one of the organization's most successful campaigns occurred in 2011 when its CEO and Founder, Scott Harrison, asked people to give up their birthday gifts and instead ask for friends and family to donate to Charity: Water. Additionally, he asked that all people fundraise or donate for the fifth "birthday" of the organization. In a campaign video, Harrison talked about all that "you" did to make this possible, particularly through physical activity:

> you biked, you ran, you walked across America, you skated and surfed, you sang, and you danced, you sold lemonade and recycled, you gave up thousands of birthdays and asked for donations instead

of gifts .... In just five years, you took a simple story and did more
than we thought possible.

(Charity: Water, 2018)

In this instance, and in many others across the water/development/charity
media landscape, the donor/fundraiser is clearly the focus.

Such messaging within water charities is significant because many of
those who donate to organizations like Charity: Water never have the
opportunity to see any type of water crisis in person. This disconnection
between donors and the realities of water insecurity makes education
around issues of water shortages and under-development particularly
important. In addition, the messaging of Charity: Water is illustrative of
the ways in which environmental issues, particularly in a charitable context, are often framed through marketing approaches rather than educational ones. People presented with the narrative of water charity on most
websites are often framed as consumers – we might think of the declaration "I shop therefore I am," popularized by artist Barbara Kruger, as
constituting our consumptive desires (Harvey, 2007, p. 170). Or, as Oprah
Winfrey once intoned about another global charitable program,
(PRODUCT)[RED], we are all "going shopping to save the world."

Such approaches to water, the environment, and development have several effects. They reify the understanding of the individual as simply
a consumer, which in turn reproduces the structures that depoliticize both
individual and collective agency (see Harvey, 2007) and create an
embodied consumer/environmentalist/donor (Bayliss, 2014). In turn, as the
consumer, "you" remain nonetheless complicit in reproducing the conditions of emergence (Butler, 2005) that give rise to those in need, given the
connections between global capitalism and global water insecurity (Glennon, 2009). Further, the resources, ability, and time needed to engage in
physical activity for charity, to fundraise, or to purchase articles of clothing in the name of philanthropy, are themselves a form and act of privilege that arguably sustain inequality more than challenge it.

Overall, in the specific context of this case study, it is important to
remember, as Bakker (2010) argues, that "water is both political and
biopolitical ... carrying with it vectors of disease and pollution, water
simultaneously connects individual bodies to the collective body politic"
(p. 190). These insights can help to frame critical considerations of the
ways in which organizations like Charity: Water approach their work,
and particularly how they recruit and educate donors and volunteers.
Charity: Water's campaigns promote the idea that global North citizens
should raise money, often through sports events like running or bicyc-
ling. What is often missing in such approaches, though, is an informed
understanding of actual environmental issues. The next section discusses
this.

Sport, education, and awareness

From the example of Charity: Water, it is clear that fundraising is a significant issue for non-governmental organizations (NGOs) working in the field of environmental sustainability. Yet while Charity: Water responds to environmental issues primarily through the logic of individual giving, which may dissociate its donors from the realities on the ground, other organizations have attempted to promote critical awareness of environmental issues among donors. Interestingly, organizations like Sport for Water Society,[3] a water development agency providing water to villages in Ethiopia, have used sport as a catalyst for such consciousness raising.

On the one hand, like Charity: Water, Sport for Water Society utilizes a road race to raise funds to build water systems in developing countries while also working with schools and development programs to teach students about water issues and to help to create curricula on water resources that are utilized in the classroom (see Bunds, 2017). In this way, Sport for Water Society seeks to help end water crises in the villages it aids by creating lifelong, educated supporters, with sport as a fulcrum for leveraging such education. In contrast to other organizations, however, Sport for Water Society sees education around issues of water security among donors as a greater impediment to the sustainability and efficacy of its organization, as opposed to simply fundraising itself. Such a concern for education reflects the EEFS approach noted above, whereby education is viewed as a key pathway to enable people to positively contribute to sustainability, not only financially, but also politically. Indeed, as the co-founder of Sport for Water Society noted, if people truly understood the serious need for water and access to it around the world, and if they could understand how water worked in their own lives, then "we" (i.e. the global North) would be less likely to allow water poverty to continue. A hurdle for NGOs like Sport for Water Society is thus how to promote environmental education for sustainability. Indeed, while Sport for Water Society is one of the more successful water development agencies and operates as a model of sorts for similar organizations in the United States, Canada, Europe, and countries in Africa, there remains a concern within the organization that it has not effectively engaged people about the seriousness, or systemic causes, of water security, and that this lack of education ultimately compromises the sustainability of the donor model itself.

In this regard, the administrators of Sport for Water Society spoke openly to me about sport being one of the options available to them to promote education, and in so doing, engage and maintain relations with donors. For example, sport events were viewed as a means by which donors might develop empathy – rather than sympathy – for the realities of water insecurity and the difficult tasks of charitable organizations while also providing opportunities to raise funds through race sponsorship

(Bunds, 2017). Educational opportunities provided through sport were specifically seen as key to such experiences because they could connect individuals across geographic and socio-economic boundaries in a way that might allow them to "be exposed to their own environmental struggles in their contexts and learn about water worldwide in a way that promotes solutions, not pity" (Bunds, 2017, p. 172).

The key point here is that the popularity, intelligibility, and fun of sport and physical activity were seen as important catalysts and mediums for promoting environmental education that could engage donors beyond a simple monetary transaction. At the same time, however, there was also a sense among members of this organization that the time and space available for engaging with potential donors in a sustained and genuine educational model was extremely limited. It was in recognition of these limits that the concept of using virtual reality technology at sports events was discussed, particularly as a way to "place" people in areas of need.

Learning about the environment through sport and immersive education technologies

The discussion above shows that charity-based environmental organizations, and particularly those working in the field of water access, desire ways to connect to and engage with individuals (and potential donors) about environmental causes. Specifically, they seek to educate people about critical issues through hands-on approaches that might cross geographic and socio-economic boundaries. With this in mind, immersive learning technologies have emerged in recent years as a potentially fruitful avenue for such engagement.

The possibilities for immersive learning technologies are profound. According to Clarke-Midura and Dede (2010): "Research on immersive environments and mediated experiences proves that one can create environments capable of capturing observations and studying authentic behaviors not possible in a conventional classroom setting" (p. 311). Such technologies might allow people to traverse (virtually) into the global South using hand-held, immersive ocular sensors (i.e., Oculus Rift) supported by Cloud- and crowd-sourced data. With respect to water issues specifically, users of such technology could gain a bird's-eye, or immersive, view of water sources via web-based geographic information systems and story maps (i.e., WebGIS and ESRI StoryMaps). These technologies could also facilitate exploration of landscapes for specific purposes, such as siting a well for drinking water, through the use of immersive 3D mapping.

A key point here is that placing individuals (including, but not limited to, aid workers, citizen scientists, museum visitors, teachers, or students) within a spatial context of water scarcity and the response thereto allows for heretofore unprecedented visualization and personalization, potentially leading to transformative educational experiences that are informative and engaging with respect to environmental issues. Given that it is neither feasible (nor desirable) for potential donors in the global North to be physically in the places of water scarcity, virtually immersive engagements can be "leveraged to influence and shape the ongoing transformation of real-world identities" (Dunleavy, Dede, & Mitchell, 2009, p. 9). The result may be the education of and engagement with individuals who see, hear, and interact with water scarcity in relatively tangible ways.

Additionally, some immersive learning technologies allow participants to interact with real environmental information through tangible landscapes by using an open source interface for 3D sketching powered by GRASS GIS. Such a system allows for the creation of 3D molds of real land masses so that individuals can learn how diseases spread, water flows, or even how building a sport facility can negatively impact the environmental landscape. This can even be coupled with an immersive virtual environment in which users can learn about their own impacts on the landscape in three dimensions.

I suggest that such technologies offer an opportunity to engage people about the environment in the context of sport. There are at least three ways in which this might be done successfully: by utilizing immersive educational technologies to educate people at sport events; by including immersive educational technologies in established sport programs; and/or by making sport the focus of immersive educational technologies. I discuss each here.

Many sporting events are comprised of multiple spectacles occurring in and around the main event. Take the marathon as an example. The day before a major marathon, there is typically an exposition or trade show in which different products are showcased while participants pick up their racing packets. Additionally, there are often groups of people at large marathons who cheer on runners and wait on family and friends. Such events provide organizers or organizations the opportunity to offer an immersive virtual reality experience. As Roussou (n.d.) has detailed, many museums, cultural centers, and entertainment venues have now turned to immersive technologies to capture public attention and promote education in relatively short periods of time.

Using such immersive technologies at a marathon to raise money for water charities would allow organizations to do more than just raise funds. It would help them to "place" people connected to the race "with" someone who has to walk long distances for fresh drinking water. It could also take runners through the maintenance of fresh water, explain how engineers devise ways to fix water delivery issues, or explore how community members create and operate solutions to water crises. It could also be used to show runners how they received their water supply for their race, and the infrastructure needed for this.

Such technologies could also be used as part of established sport for development programs, particularly ones focused on youth and life skills. In my city – Raleigh, North Carolina – a partnership between Raleigh, the Capitol City Crew, North Carolina State University, and the Carolina Hurricanes hockey team provides youth at boys' and girls' clubs with "learn to play hockey" lessons, followed by life skills. Youth spend 45 minutes on and 45 minutes off the ice. In such a program, it would be possible to engage students in immersive technologies, online landscapes, or a number of other virtual experiences that would allow them to engage with environmental issues in a tangible and immersive way.

Finally, considering the importance of environmental education, it is reasonable to suggest that the popularity of sport could be used to draw attention to environmental issues and the importance of sustainability. Specifically, since sport is known to be detrimental to the environment on a regular basis via waste, air pollution, land degradation, water usage, and energy usage, among other issues (Bunds & Casper, 2018), immersive technology might offer a way to engage with and educate the estimated 50 million youth sport participants in the United States alone (Bunds et al., 2018), particularly about the ways in which sport impacts the environment and their responsibility to such issues. From this perspective, immersive technologies offer an opportunity to promote and disseminate understandings of the connections between sport, the environment, and sustainable development.

Conclusion

Of course, utilizing immersive educational technologies in the field of sport as a way to teach individuals about the environment will not automatically change behavior, solve environmental issues, or even mitigate genuine concerns about the limits and politics of consumer philanthropy. However, such tools can be utilized to provide individuals with information they need for informed engagement, rather than simply rendering them the object of marketing and fundraising campaigns. As Tilbury (1995) argues, in order to "fix" issues related to environmental sustainability, there must be broader and more precise understandings of environmental issues and solutions. To that end, organizations might utilize immersive technologies, in and through sport programs, as a way to educate and inform about water in ways that challenge the notion of "you" as a savior. In turn, rather than ending with an "ask" for donations, such an approach might encourage thinking and learning about genuine, sustainable solutions to mitigating water crises.

Overall, using sport as a hook to educate people in an EEFS model and engaging them with immersive technologies that create visceral experiences are potentially untapped approaches to the challenge of placing sport in the service of environmental sustainability.

Notes

1 I was first attracted to this area after reading the work of Donnelly et al. (2007) in their report on the use of sport to foster child and youth development and education and have carried this through with my work utilizing immersive education technologies.
2 "London Water Charity" is a pseudonym.
3 "Sport for Water Society" is a pseudonym.

References

Bakker, K. (2010). *Privatizing water: Governance failure and the world's urban water crisis*. Ithaca, NY: Cornell University Press.

Bayliss, K. (2014). The financialization of water. *Review of Radical Political Economics*, 46(3), 292–307.

Bunds, K. S. (2017). *Sport, politics, and the charity industry: Running for water*. New York: Routledge.

Bunds, K. S. & Casper, J. M. (2018). Sport, physical culture, and the environment: An introduction. *Sociology of Sport Journal*, 35(1), 1–7.

Bunds, K. S., Kanters, M. A., Venditti, R. A., Rajagopalan, N., Casper, J. M., & Carlton, T. A. (2018). Organized youth sports and commuting behavior: The environmental impact of decentralized community sport facilities. *Transportation Research Part D: Transport and Environment*, 65, 387–395.

Butler, J. (2005). *Giving an account to oneself*. New York: Fordham University Press.

Charity: Water. (2018). September campaign 2011 trailer. Retrieved on July 17, 2019 from: www.charitywater.org/stories/videos/watch?v=1453

Clarke-Midura, J. & Dede, C. (2010). Assessment, technology, and change. *Journal of Research on Technology in Education*, 42(3), 309–328.

Coalter, F. (2007). *A wider social role for sport: Who's keeping the score?* New York: Routledge.

Darnell, S. C. (2007). Playing with race: Right to play and the production of whiteness in "development through sport". *Sport in Society: Cultures, Commerce, Media, Politics*, 10(4), 560–579.

Donnelly, P., Darnell, S., Wells, S., & Coakley, J. (2007). The use of sport to foster child and youth development and education. In Sport for Development and Peace International Working Group, *Literature reviews on Sport for Development and Peace*, 7–47. Toronto, Canada: Sport for Development and Peace International Working Group.

Dunleavy, M., Dede, C., & Mitchell, R. (2009). Affordances and limitations of immersive participatory augmented reality simulations for teaching and learning. *Journal of Science Education and Technology*, 18(1), 7–22.

Edwards, M. B. (2015). The role of sport in community capacity building: An examination of sport for development research and practice. *Sport Management Review*, 18(1), 6–19.

Glennon, R. J. (2009). *Unquenchable: America's water crisis and what to do about it.* Washington, DC: Island Press.

Harvey, D. (2007). *A brief history of neoliberalism.* New York: Oxford University Press.

Hayhurst, L. M. C. (2011). *"Governing" the "girl effect" through sport, gender and development? Postcolonial girlhoods, constellations of aid and global corporate social engagement.* Ann Arbor, MI: Proquest.

IUCN/UNEP/WWF. (1980). *World conservation strategy: living resources for sustainable development.* Gland, Switzerland: IUCN/UNEP/WWF.

IUCN/UNEP/WWF. (1991). *Caring for the Earth: A strategy for sustainable living.* London: Earthscan Publications.

Jones, G. J., Edwards, M. B., Bocarro, J. N., Bunds, K. S., & Smith, J. W. (2018a). Leveraging community sport organizations to promote community capacity: Strategic outcomes, challenges, and theoretical considerations. *Sport Management Review*. 10.1016/j.smr.2017.07.006.

Jones, G. J., Wagner, C. E., Bunds, K. S., Edwards, M. B., & Bocarro, J. N. (2018b). Examining the environmental characteristics of shared leadership in a sport-for-development organization. *Journal of Sport Management*, 32(2), 82–95.

Kaufman, P. & Wolff, E. A. (2010). Playing and protesting: Sport as a vehicle for social change. *Journal of Sport and Social Issues*, 34, 154–175.

NSF. (2017). Advancing informal STEM learning (AISL). Retrieved on April 14, 2018 from: www.nsf.gov/pubs/2017/nsf17573/nsf17573.htm

Roussou, M. (n.d.). *Immersive interactive virtual reality in the museum.* Retrieved on July 17, 2019 from: https://pdfs.semanticscholar.org/7493/6f7e348e5af66f556441b1723b336b58fea7.pdf

Saad, L. (2017, March 27). Half in U.S. are now concerned global warming believers. Retrieved on June 1, 2017 from: http://news.gallup.com/poll/207119/half-concerned-global-warming-believers.aspx

Sage, G. H. (1990). *Power and ideology in American sport: A critical perspective.* Champaign, IL: Human Kinetics Books.

Sauve, L. (1996). Environmental education and sustainable development: A further appraisal. *Canadian Journal of Environmental Education*, 1, 7–34.

Skinner, J., Zakus, D., & Cowell, J. (2008). Development through sport: Building social capital in disadvantaged communities. *Sport Management Review*, 11, 253–275.

Spaaij, R., Farquharson, K., & Marjoribanks, T. (2015). Sport and social inequalities. *Sociology Compass*, 9(5), 400–411.

Spaaij, R. & Jeane, R. (2013). Education for social change? A Freirean critique of sport and development and peace. *Physical Education and Sport Pedagogy*, 18(4), 442–457.

Stevenson, K. T., Peterson, M. N., & Bradshaw, A. (2016). How climate change beliefs among US teachers do and do not translate to students. *PLOS One*, 11(9), e0161462.

Tilbury, D. (1993). *Environmental education: Developing a model for initial teacher education.* PhD thesis, University of Cambridge.

Tilbury, D. (1995). Environmental education for sustainability: Defining the new focus of environmental education in the 1990s. *Environmental Education Research*, *1*(2), 195–212.

Woodhouse, J. L. & Knapp, C. E. (2000). *Place-based curriculum and instruction: Outdoor environmental education approaches. ERIC Digest*. Retrieved on July 17, 2019 from: https://files.eric.ed.gov/fulltext/ED448012.pdf

Wright, A. & Coté, J. (2003). A retrospective analysis of leadership development through sport. *The Sport Psychologist*, *17*, 268–291.

Chapter 5

Postcolonialism, pristine natures, and producing the village green

Pastoral Englishness in Indian cricket-themed gated communities

Devra Waldman and Gavin Weedon

Introduction

Cricket evokes a pastoral imaginary of England past: a rural nostalgia centred on luscious greens, and in older forms of the game, long temporalities that stretch matches over days and weeks rather than seconds and minutes. Tropes of fields surrounded by elms, observing games from woodland edge, picturesque hedges framing open pavilions and fields surrounded by barns and mills recur across literary descriptions of the game (Bale, 1994; Ross, 1981). In the much-loved satirical novel about village cricket written in 1935 by A. G. Macdonell, *England, Their England*, the cricket ground is positioned as the anchor of an imagined England that is 'unspoilt by factories, financiers and tourists' (Macdonell, 1935, p. 107). Such visions of nature, landscape, and rurality persist in representations of (English) cricket, as if a premodern paradise lost was somehow preserved in this cultural form. John Bale (1994) makes this point in observing that:

> the landscape of cricket, elicited from writers, painters, and others ... permit[s] the construction of a simple model of the English cricket landscape ensemble ... [that] contain a number of distinct elements. Trees, shrubs, church, barn ... the field itself, the immaculately maintained 22 yard wicket contrasting with the greener outfield.
>
> (p. 155)

Of course, the making and maintaining of these pristine greens and immaculate wickets is a matter of much significance in the outcome of a match, an observation that gestures to the work involved in producing a wicket, a green, a field: in sum, the pastoral image of cricket itself.

This image is integral to the colonial history of cricket: the story of a game ensconced in a long-sedimented aura of Englishness being taken to far-flung fields to subjects who, in most of this storying, subsequently react to the imposition of this quintessential cultural form. A colonial history of this arc is well established, its veracity is seemingly attested by the regions in which cricket remains popular (see Malcolm, 2012). And yet, if we consider this history and its attendant imagery with a postcolonial sensibility, there is cause to doubt both the association of any distinctive Englishness with cricket (Carrington & McDonald, 2001) and the extent to which the England denoted in such imaginaries has ever existed (Williams, 1973). Postcolonial thought, notwithstanding its diversity, aids in understanding the legacy of colonial histories and in critically interrogating and recasting those histories from the first encounters to the present. With such a sensibility, Thomas Fletcher (2011, p. 21) asserts that 'the story of imperial cricket is really about the colonial quest for identity in the face of the colonizers' search for authority'. Note here the proposition that Englishness is produced *through colonial encounters* rather than existing fully formed prior to its exportation. On this basis, Mike Marqusee (1994) has argued that 'one of the things that makes English cricket *English*' is, in fact, 'the way it lies about itself to itself', adding that '*The Englishness is in the lie*, in the cult of the honest yeomen and the village green, in the denial of cricket's origins in commerce, politics, patronage, and an urban society' (Marqusee, 1994, p. 61, original emphasis). The argument emerging here is that the game of cricket did not simply diffuse out of England and land intact 'somewhere else' to be assimilated or resisted, but was part of the making of Englishness as an image, working to materialize a certain sense of identity, history and culture in both colonizer and colonized. This is the staging of cricket as an image (cf. Mitchell, 2000), and central to this process was a sense of a pure and venerable Nature that is inseparable from the cricketing ideal, the greens, the whites, and all the rest.

This chapter centres on just such an image, albeit one that appears on first glance to have twenty-first-century origins. In 2011, the Marylebone Cricket Club (MCC) – the highly exclusive private members club established in 1787 that operates out of Lord's Cricket Ground (the 'home of cricket') in London, England – entered a brand licensing agreement with an international real estate investment company based in the United Kingdom called Anglo Indian. With exclusive rights to the MCC and Lord's Cricket Ground brands, Anglo Indian propose to develop 12 branded, gated communities in major metropolitan regions of India that simulate iconic elements of the MCC's and Lord's architecture. Each community would be built around a duplicated Lord's Cricket Ground and MCC private members club, and feature an MCC cricket academy, a Lord's tavern-themed restaurant, Lord's merchandise outlets, and residential facilities. All of this is aimed at the Indian upper middle classes: according to at least one commentator, the 'last living Englishmen' through whom (the appeal of) colonial Englishness survives the formal fall of empire (Mudderidge, quoted in Appadurai, 2015, p. 1).

Many issues arise from these proposals, not least those concerning the (re)colonization of land by sporting institutions ensconced in an aura of Englishness together with British-based real estate companies courting Indian middle-class consumers (Waldman et al., 2017). Our focus in what follows is on the 'imagineering' of pastoral Englishness in these cricket-themed oases, which forms the material, symbolic, and affective anchor point for exclusive community building. We begin by considering how myths of rarefied and pristine naturality are part of the logic of modernity, and inseparable from colonialism in their 'othering' of nature. With a postcolonial sensibility, we then turn to the MCC-Anglo Indian proposals and critically interrogate their proposals for enclosed, Anglicized and cricket-themed communities in a former British colony. The contention throughout is that the 'purification' and 'othering' of nature is inseparable from other logics of appropriation, exclusion, and domination (i.e. axes of race, class, sex, and colonization; Plumwood, 1993), as evidenced by these proposals for urban sport-related developments in postcolonial India. We suggest, by way of conclusion, that studies of 'the environment' or sustainability in sport-related contexts consider (post)colonial histories of sport in relation to the constructions, representations, and productions of nature.

Myths of nature: modernity and purification

Critical analysis of the production of pristine nature in any cricketing space might begin and end with the observation that what is produced is a fallacy: the cricket green, the field, the England it is intended to evoke, are mythologies of scenes and places that never existed in such rarefied forms. Such an analysis might proceed as follows. Despite the most common presentations of cricket landscapes as 'nature unspoilt', cricket grounds have long undergone scientific preparation, careful manipulation, and geometrical arrangement with measurements of wickets, stumps, and creases (Bale, 1994). In 1864, the first groundsman was employed at Lord's Cricket Ground, where the lines between the stadium and the garden began to be blurred through the expertise of those who tended to them (Cole, 1982). The groundsmen of cricket clubs in late nineteenth-century England were also *gardeners*, and were thus intimately familiar with horticulture (Bale, 1994). Percy Peake, employed at Lord's Cricket Ground in 1874, 'had been a gardener and keeper of lawns before ... he was an expert in agriculture ... his careful and empirical study of soils, grasses and marls gradually produced a surface at Lord's that became the envy of every county club' (Sandiford, 1984, p. 278). It was then that the landscape of cricket and the rise of the gardener-groundsman began to produce a field of perfection, smooth playing surfaces, and practices of turf management (Bale, 1994). When confronted with the labour, discourse, and technology required to produce and sustain these spaces, we can see that 'cricket's apparent naturalness is a myth' (Bale, 1994, p. 162).

Documenting the 'ideological abuse' (Barthes, 2000 [1972], p. 11) that underlies the mythology of cricket grounds as pristine natures doubtless has its merits. Yet there is a danger that exposing the 'falseness' of the cricket ground's natural aesthetic would imply its purity hitherto, as if cricket did indeed take place in England's green and pleasant lands before the onset of industrialism, the 'interference' of technology, and the representational strategies of media. Put another way, it is important to be mindful of dualisms in our approach not only because of the analytical culs-de-sac to which they often lead, but because of the established histories and identities that such dualisms recapitulate. For just as a harmonious balance between (human) culture and 'the environment' is implied by the topographical structure of cricket, and indeed most forms of modern sport, comparable nature–culture dualisms *are produced through* the social and historical conditions in which this cultural form has been shaped.

To unpack this contention, Bruno Latour (1993) has advanced an influential thesis on what he calls the 'purifying' logic of modernity that helps us begin to explain the persistence of myths of nature. In broad strokes, Latour's thesis is that modernity – defined variously by diverse commentators as epoch, mood, condition, experience, project, and aesthetic – can be understood through its dual organizing logic of purification and translation. Through purification, the world is divided into two different, incommensurable realms: nature and culture. The first is a realm of timeless things, immutable in nature and awaiting representation; the second, a realm of turbulent human activity, of culture, politics and communication. What makes 'us' modern is not that we live in the modern epoch, as this claim to historical rupture has always accompanied the language and concept of modernity (Gruneau, 2017; Kumar, 2005). Rather, it is to believe that 'we' moderns occupy the latter category of culture, having transcended nature, as well as our ancestors in prior, premodern societies. Rather than take sides, Latour's contention is that this dualism is a fallacy: 'We have never been modern', have never and could never pry the world asunder into divergent realms cleansed of one another. And yet, through the performative power of the belief that 'we' have transcended our animal origins, our premodern ancestors, and our dependence on the environment, 'we' have weaved these morphisms of nature and culture together at unprecedented rate and scale.

Through the bifurcating modern logic of purification, it is easy to see why pristine nature, signifying purity and simplicity, carries such great economic and cultural appeal. Pastoral nature becomes enchanting in a world where 'it' is increasingly hard to find, and difficult to produce. It is also possible to see why modern sport is such a seductive vector for these myths. As sport is often heralded as a bastion of human achievement, a trial of character, strength, and control of one's senses, it is often cast against a backdrop of nature (elements, landscapes, one's own physiology) that serves as a site for such achievements. For example, in his work on the political ecology of skiing, Mark Stoddart (2012) has shown how the 'mountainous sublime' is crucial to the allure of ski tourism. The 'mountainous sublime' – think of snowy peaks adorned with fresh, unblemished powder awaiting the skiers' descent – is the quality of pristine naturality desired by many skiing aficionados. To meet these desires and the interests they serve, this quality must be *produced* through a socio-technical imbroglio of 'actors' including snow machines, ski chairs, magazines, volunteers, highways, and signage, all mobilized in service of sustaining skiing's natural veneer. Note Latour's dual processes at work here: the purification of mountain ranges in order to produce and sustain the mountainous sublime itself necessitates the creation of an unprecedented number of 'hybrids', at once natural, cultural, and technological, in order to promulgate the myth of purity for human pleasure and, of course, for profit. Similarly, Millington and Wilson (2016) write of Augusta National Syndrome – the visual emergence of Augusta's meticulously manicured golf courses for television audiences since the 1960s – as a key development in the 'greening of golf' through chemical treatment and, latterly, environmental activism and legislation. These analyses each demonstrate why dualisms are as pervasive and profitable as they are misleading. Perhaps their most significant shared contention, though, is that these 'pristine natures' are made – produced as material, symbolic, and discursive realities – and that their making stems from complex cultural, social, and economic interminglings and imaginaries.

(Post)colonialism, colonial modernity, and nature as 'other'

Two crucial qualifications are needed to this account of modernity as a purifying logic. First, as Val Plumwood (1993) shows in her ecofeminist critique of the mastery of nature, the nature–culture dualism does not exist in silos. It is structurally linked to other modes of domination. The purifying of nature 'results from a certain kind of denied dependency on a subordinated other' (Plumwood, 1993, p. 41). In this schema, that which is deemed to be Of Nature is marked as 'other' in opposition to 'me' and in different contexts is either tamed, feminized, domesticated, or tempered, and above all subject to reason. Only then might the inhabitant of the Master category be concerned with what could be called sustainability, or the heroic saviour narrative, for conservation always follows from (perceived) domination (Haraway, 1989). This mode of subordination and denied dependency on the basis of otherness links nature–culture as dualism to many modes of oppression including gender, class, race, species, and colonization. Think, for example, of the subordination of women on the basis of claims to biological limits; of the oppression of non-white Europeans on the basis of 'closeness to nature' and associated 'under-development'; of the reduction of nonhuman animals to nature and the consumption of meat as a masculine enterprise; and the uneasy casting of the planet as 'mother earth' alongside ongoing environmental degradation. Now consider the denied dependencies implied by each of these, such as the role of women's social reproductive labour in the history of (at least) capitalist production; of the non-European material basis of wealth and privilege produced through colonialism; of the reliance on land, air, water for all of life on earth.

Plumwood's ecofeminism is joined by many strains of postcolonial theory that also observe the postulation of culture and reason as separate and superior to nature as part of the logic of colonization. Indeed, Euro-American modernity has often measured its progress on the basis of how far it is deemed to have travelled from nature (Anderson, 2007; Gregory, 2001). The points to emphasize here – there are many whose complexity escapes these broad strokes – are that the purification of nature is fore-most in the colonial imaginary, which survives the end of formal colonial-ism and continues to pattern modern European imaginaries, not least in the struggle to accept nature into the rational domain of politics, and that purification of nature cannot be untethered from other modes of oppres-sion and the dependent relations they belie. Framing 'the environment' as its own domain is therefore not only inaccurate, but ethically fraught inso-far as this performs an erasure of the dualistic relations to which the puri-fication of nature is tethered.

The second point is that 'nature' as a rarefied, idealized realm is not a product made and exported by Europeans (despite modernity often acting as a synonym for the West). This second point might seem to somewhat contradict the first: if European modernity has incubated a dualistic logic of nature–culture that informed the logic of coloniza-tion, then how is Europe *not* the origin of this logic and its repercus-sions? To this we emphasize the significance of the *colonial encounter* as productive in its own right and involving all of those implicated, as opposed to accepting the supreme efficacy of the *colonial conquest* in which those cast outside of the Master category simply acquiesce to a preconceived, fully formed force of European identity distinguished by reason and culture (Bhabha, 1994). To such Eurocentric histories, especially those that herald a 'temperate environment' in Northern Europe as a determinant of colonial influence and economic growth (Braut, 1999), postcolonial and decolonial sensibilities help to recast established accounts of European modernity and colonialism. Enrique Dussel (1995), for example, identifies in modernity and colonialism a shared birth in Columbus's voyage to the Americas in 1492 that proved mutually reinforcing thereafter. Homi Bhabha (1994) contends that colonial power should be understood as a production of hybridization, rather than as simply the efficacious commands of the colonizer and/or silent repression of local traditions. Every concept/institution that the colonizer brings to the colonized will be reborn, renewed, and reinterpreted in light of the local culture. This emphasis on hybridity troubles the presumption of colonial domination over colonized populations (Bhabha, 1994). Elsewhere, Ann Stoler (1989) offers an account of Dutch settlers in the East Indies forging a collective, racialized identity through the colonial encounter based on class and ethnic distinctiveness, from which a bourgeoisie European identity subsequently took shape in the metropole. Timothy Mitchell (2000) recounts how the emergence of the word 'nationalism' in English was popularized in the mid-nineteenth century only after the proliferation the word 'international', notably in London's Great International Exhibition and Karl Marx's Working Men's International Association.

Each of these dispersed accounts of colonialism cast into question the diffusion of secure national imaginaries from metropole societies into colonies. In turn, they invite us to consider the production of pristine nature in cricket not as a movement from England to India, but as the product of this encounter. Indeed, it is often remarked that cricket now looks more Indian than English to the postcolonial populations on the Subcontinent (Nandy, 2000).

Anglo Indian cricketing 'paradise': a postcolonial development?

The postcolonial and its relation to colonialism often turns on memory and its absence, or nostalgia and amnesia: 'the terrible twins', in Terry Eagleton's words, characterized by 'the inability to remember and the incapacity to do anything else' (quoted in Gregory, 2004). This acerbic observation too often rings true where colonialism is either denied, selectively celebrated, or lived without reflexivity in settler-colonial societies. The ways in which England 'lies about itself to itself' through cricket (Marqusee, 1994), as a land of leisurely classes playing at a quintessentially English game against the backdrop of venerable nature, is exemplary of this point. It is therefore both curious and hardly surprising that this same cricketing ideal is at the heart of the MCC and Anglo Indian development plans for gated communities in India. What has changed, and what has stayed the same, in order for this imagination of Englishness to carry cultural, social, and commercial appeal in twenty-first-century India? Focusing this line of inquiry, what role does a purified, Anglicized vision of pastoral nature play in these proposals for cricket-themed gated communities? These questions are our focus for the remainder of the chapter, where we offer suggestions for those who are looking to study sport-environment-development with a postcolonial sensibility.

In a marketing brochure aimed at would-be inhabitants, these developers cast their vision of 'paradise':

> Picture a living, breathing community built around a nation's biggest passion … our vision builds a cricket enthusiast's idea of *paradise*. Whether you are a visiting business executive, a school teacher or a family seeking entertainment and relaxation at the weekend, Anglo Indian developments will provide you with a holistic experience for the entire family ….
>
> (Anglo Indian, 2014, p. 31, our emphasis)

This 'imagineering' (Silk, 2004) of unspecified spaces in urban cities blends mythical ideals of nature with the promise of expansive green space for leisure and relaxation. In the proposed layout of the communities, the simulated Lord's cricket ground is the centre and focal point of the community. Immediately surrounding the cricket pitch is the duplicated MCC members' club, Lord's tavern, and retail and residential facilities. Moving concentrically outwards are marked parks, an 'eco village', an equestrian centre, and walking trails.

The spatial arrangement, then, is premised on the centrifugal aura of the Lord's cricket pitch – which is characterized by its carefully gardened, immaculate, and smoothed playing surface, meticulously maintained by those intimately familiar with horticulture, agriculture, and geology (Bale, 1994). The surrounding equestrian facilities, park spaces, and walking trails, and the rendered but unspecified green spaces that lie beyond, are all cast in this affective register: 'Like a classic English village, the cricket green provides the focal point – with a classic pavilion providing an iconic reminder of all things MCC and Lord's. But this is no style-based façade' (Anglo Indian, 2014, p. 29).

These comments and illustrations make clear that appeals to pastoral paradise and temperate greenery are inseparable from mythic versions of distinctly English (colonial) cricket. It is here that the relationships between colonialism and the imagination of nature and landscape are materialized in unspecified Indian land.

It is tempting to analyse these proposals as a simple marketing ploy on the part of the MCC and Anglo Indian, one in which the 'home of cricket' is outsourced and experienced at a profit. Yet to understand why they appear in this form, at this moment, requires a careful postcolonial reading of the history of cricket, the sports place in the colonial project, and the contemporary context of professional cricket in India. Cricket was transported to India in the late 1700s by the British who moved abroad as part of the British East India Company. While these British colonialists established the Calcutta Cricket Club in 1792 (the second oldest club in the world behind the MCC), cricket spaces were reserved for British citizens in India for almost a century. This changed in 1835 when the British Raj introduced a new policy of Anglicization, whereby English-educated Indian elite were to be trained in English tastes, manners, and customs in order to assist in colonial bureaucracy and become models for other Indian citizens (Appadurai, 1996; Cashman, 1998). While there was no formal initiative in place to utilize sport as a means to support the Raj, cricket quickly became part of the British 'civilizing' process, with many British officials proving to be enthusiastic missionaries for the introduction of cricket under this policy (Appadurai, 1996; Mills & Dimeo, 2003). To these missionaries cricket was an appropriate choice to assist in Anglicization; to know, understand, and play cricket in particular manners was also to know, understand, and embody a preferred Englishness (Malcolm, 2001, 2012).

Soon after, Parsis of Bombay – an elite community that developed close links with the British in the interests of increasing their status – took interest in the sport (Appadurai, 1996; Cashman, 1998). Indian princes also embraced the playing and development of cricket skills as they perceived the game to be an extension of already existing local traditions, a means to provide new spectacles for their subjects, and providing an opportunity to open up new doors for economic profitability in England (Appadurai, 1996). The princes brought in English and Australian professional cricketers to teach and train their own teams, as well as providing financial support to talented players from lower caste and class backgrounds so they could develop the game and their skills (Appadurai, 1996). It was this hierarchical cross-hatching of British gentlemen in India, Indian princes, mobile Indian men who were part of civil services under the Raj, and white cricketing professionals who trained the first great Indian cricketers for the first decades of the nineteenth century that laid the groundwork for the contemporary 'Indianization' of cricket (Appadurai, 1996; Bose, 2006). To Nandy's (2000) witticism that 'cricket is an Indian game accidentally discovered by the English', we observe here the process of hybridization (Bhabha, 1994) as the game was not simply imposed, but taken up and reshaped in the intermingling of British and Indian cultural influences.

Cricket has thus been remade to be central to distinctive Indian national identity, with India having emerged as the cultural and economic centre of the sport (Gupta, 2009). India represents 80% of the income for the International Cricket Council, ensures more test matches take place in South Asia, and has the incredibly successful and lucrative professional league (the Indian Premier League, IPL) that attracts many top players from around the world (Majumdar, 2007; Mehta et al., 2009). In its inaugural year, the IPL cleared a profit of $1.8 billion, with a viewership of 220 million (and growing to over 320 million in 2017) within India, demonstrating the sport's incredible popularity and profitability in the domestic market alone (Gupta, 2009; Mehta et al., 2009).

This makes Anglo Indian's decision to build these community around the MCC and Lord's Cricket Grounds brands especially curious. Historically, as a highly exclusive club based in London, many members of the MCC were also part of the political classes, aristocracy, or social elite, leading some to argue that the cricketing and political British Empire were one and the same (Stoddart & Sandiford, 1998). These brands are bastions of a British colonial past, and the symbolic and aesthetic importance of the MCC/Lord's as particular English (colonial) institutions is central to the imagining of spaces. Implied through the assertion that residents will constantly be surrounded by 'reminders of all things MCC and Lord's', these developments will be 'quintessentially English', looking to replicate the rightful 'home of cricket', while individuals to whom this appeals can bask in memorabilia of 'bygone sportsmen' (p. 22). Yet the end of formal colonialism and the economic dominance of India in professional cricket recast these development proposals as a perverse recognition of England's waning importance and power within global cricket, and an attempt to reassert dominance lost through building spaces with constant 'reminders' of English cricket history and heritage. No longer able to claim or assert economic dominance, the MCC is trading on a sense of Englishness that retains an appeal among wealthy Indians. Pastoral nature of the cricket green and broader spaces that surround it is as crucial these proposals and their commercial success as it is to ideas of the sport of cricket.

Questions also arise about the kind of work required to produce and sustain these versions of 'natural' space, and the kinds of exclusions and modes of subordination and denied dependency to which they may be tethered. The cricket ground here (in its mythologized English form) is not necessarily the context for the performance of English identities, but more the scene for the purification of space – which is intimately tied to colonial processes whereby nature is both dominated and domesticated (Gregory, 2001; Mitchell, 2000). Following Latour (1993), the bifurcating logic of purification is visible and reproduced through the networks required to produce and maintain these pristine and seemingly natural spaces. Particularly, the landscape will be cleansed in order to produce and sustain the image of the manicured greens, groomed paths, and immaculate parks. These spaces will all be 'hybrids' – the result of multiple combinations of natural, technological, and cultural components to sustain the myths of the purity of the cricket green and grounds that surround it. Enframings (Gregory, 2001) of nature unspoilt underpin the emphasis on a seemingly naturalized idyllic environment in these proposed developments. In particular, it is the distinctiveness of a highly stylized

English cricket green with the smell of freshly cut grass, walking on the manicured and flowered paths, or lounging in an open park – all of which will be maintained through aesthetic regimes that will be strictly enforced to police space (Waldman et al., 2017). Understandings of the environment become framed as 'improved nature', no longer wilderness, but instead immaculate green lawns, flower beds, pruned trees on which to be capitalized (Baviskar, 2011).

It is important to consider that although these enclaves are conceptualized as idyllic and open, they will have gates and walls that will continually work to police the boundaries of those who are who are allowed in and those who are not and under what conditions. Buying and living in these exclusive and serene environments is an 'opportunity' for individuals to display a distinct and exclusive lifestyle within a controlled and secure space: 'By "Invitation Only" – the Lord's Club, India: Certain timeless qualities define the essence of a classic private members club, and our new clubs will have them all … with all the privacy and amenities one would expect …' (Anglo Indian, 2013).

The imaginations of space claim to celebrate and encourage diversity, yet simultaneously the aestheticized designs and visualizations are driven by highly selective ideas about who is allowed to be and move within them (Pow, 2009). As such, these developments could be part of what Baviskar (2011) terms bourgeois environmentalist projects: the pursuit of the ordering and cleansing of nature for the use of the wealthy, and the reservation of these 'clean' spaces for those who 'properly belong'. These communities are being sold as a safe haven of greenery and serenity close to, yet distinct from, the dirty, chaotic urban cores of major the metropolitan urban areas of India (Borsdorf & Hidalgo, 2008). While beyond the scope of this chapter, the proliferation of gated, branded, residential spaces such as these is part of broader practices of spatial fragmentation in Indian cities, where increasing spaces are geared towards producing environments that satisfy the desires for the middle- and upper-class Indians for exclusive living spaces that have marked separation from the marked 'Other' in the urban core (Fernandes, 2004). Such communities that promise serene paradise are especially centred on ideas of security from violent and dangerous 'Other', the provision of beautified environments that are tranquil and quiet, and the opportunity to demonstrate a cosmopolitan lifestyle through consumption (Andrews et al., 2014; Waldman et al., 2017).

Of course, gates and walls notwithstanding, the boundaries of these communities will be permeable, as those who do not 'properly' belong will build and maintain the facilities and leisure spaces in the communities. The distinction between inclusion and exclusion is indeed blurry due to the servile work (the construction staff, maids, security personnel, grounds keepers, cooks, among others) that the privileged middle- and upper-class consumers rely on for the building, maintenance, and nurturing of these environments and the lifestyle they promise to produce. Perhaps this reflects the 'necessity' to keep the 'Other' close at hand to provide work and services, yet far away as humans – on the inside, but not included (Appadurai, 2000). Despite this, a very particular sense of place is embedded in the spatial imaginary of these branded, gated communities that speaks to privilege and hierarchies of legitimate belonging (Waldman et al., 2017).

'Never was and never will be': reflections on development imaginaries and directions for sport-environment-development research

To paraphrase Timothy Mitchell's (2000) words on modernity, we might say that the cricket ground in India is not so much the stage for performing Englishness so much as its staging; less the place of cricket's second birth so much as the scene of its purification. Though the MCC is predictably keen to situate its own distinctive brand identity in the architecture and symbolism of these developments, it is the pastoral imaginary of cricket that defines them.

Much of what is currently being written about nature and the 'environment' in the social sciences and humanities might take our analysis as exemplary of an ostensibly outmoded, representational, constructivist rendering. We might even stand accused of perpetuating the object of our own critique: reifying the purified form of nature in these gated communities rather than focusing on their materiality. Likewise, modes of political economy might favour a focus on where exactly funds are coming from and forecast to go in the production of these developments and the land they would shape. Our analysis seems not only warranted, but apposite, however, given that as we write, these developments exist only as proposals: urban imagineerings materialized in brochures such as those quoted above. Delays and complications since their 2011 proposal might even mean that, like the nostalgic visions of England to which they appeal and on which they trade, these gated communities will never exist as depicted.

Postcolonial theory offers a great deal for understanding the production of nature in sporting and sport-related spaces – indeed, a great deal more than this chapter could contain. We wish to conclude by making the case for studying 'the environment', sport, and development together with the concepts and perspectives we have worked with here. Following Carrington (2015), postcolonial insights cannot and should not be limited to 'postcolonial contexts' and examinations of particular manifestations of European myths. As Bhambra (2014) argues, a postcolonial approach entails a necessary account for narratives that are missing from dominant ones and attention to various inequalities with which they are intimately intertwined. It is important and essential, then, in a move towards a postcolonial analysis of sport, the environment, and 'sustainable' development, that research more thoroughly examine the situated and historical construction of nature that is to be sustained, conserved, or 'improved'. As discussed above, in cricket this gesture to nature is tethered to English-ness, to a venerable version of premodernity, to the leisurely classes and games pitched to the pace of the good life 'before'.

What of the many instances of sport meeting environmental issues that are perhaps not explicitly rooted in colonial imaginaries or driven so clearly by pastoral nostalgia in postcolonial contexts? If environmental issues are tethered to all the rest, as Val Plumwood's ecofeminism evinced nearly 30 years ago, then the making of wilderness or conservation parks or leisure spaces needs to be foregrounded as more-than-environmental issues. Cricket-themed gated communities, we suggest, are pronounced examples of the purification of nature, and their roots in colonial imaginaries of the environment are also overt. The obvious critique to be made here is of bourgeois environmentalism (Baviskar, 2011) and sustainability projects that are bound up with gentrification processes (Gould & Lewis, 2017). Yet, in addition to this historical, material call to interrogate the making of pristine natures, there is also an epistemological point to emphasize concerning the selection of one's object of analysis, insofar as this might be based on preexisting 'enfram-ings' (Gregory, 2001) of nature or 'the environment'. To paraphrase, the ideas of nature that persist in our environmental imaginaries may motivate our decisions about what to study and how to study it. This is a foremost challenge for critical sport-environment-development scholars to heed, one that necessitates asking how 'nature' has emerged as a category of critical analysis in a given (sporting) scene. Colonial encounters are, we argue, part of the making of pristine natures; a material-discursive entanglement that often takes shape when a rarefied natural realm requires sustaining, conserving, or saving. This might be true, as environmental crisis is real and ongoing. But unless approaches to sustainability or development recognize their 'denied dependencies' with other logics of exclusion and domination, then sustainability will remain part of the colonial bourgeois lexicon.

References

Anderson, K. (2007). *Race and the Crisis of Humanism.* London: Routledge.
Andrews, D. L., Batts, C., & Silk, M. (2014). Sport, glocalization and the new Indian middle class. *International Journal of Cultural Studies, 17*(3), 259–276.
Anglo Indian. (2013). Anglo Indian branded developments in India. Retrieved from: www.angloindian.net.

Anglo Indian. (2014). *Communities Built around Cricket: Anglo Indian Branded Developments in India*. London: Anglo Indian.

Appadurai, A. (1996). *Modernity at Large: Cultural Dimensions of Globalization*. Minneapolis, MN: University of Minnesota Press.

Appadurai, A. (2000). Spectral housing and urban cleansing: Notes on millennial Mumbai. *Public Culture, 12*(3), 627–651.

Appadurai, A. (2015). Playing with modernity: The decolonization of Indian cricket. *Altre Modernità, 14*, 1–24.

Bale, J. (1994). *Landscapes of Modern Sport*. Leicester, UK: Leicester University Press.

Barthes, R. (2000 [1972]). *Mythologies*, trans. A. Levers. London: Vintage.

Baviskar, A. (2011). Cows, cars, and cycle rickshaws: Bourgeois environmentalists and the battle for Delhi's streets. In A. Baviskar & R. Ray (Eds.), *Elite and Everyman: The Cultural Politics of the Indian Middle Class* (pp. 391–418). Delhi, India: Routledge.

Bhabha, H. (1994). *The Location of Culture*. London: Routledge.

Bhambra, G. K. (2014). Postcolonial entanglements. *Postcolonial Studies, 17*(4), 418–421.

Borsdorf, A., & Hidalgo, R. (2008). New dimensions of social exclusion in Latin America: From gated communities to gated cities, the case of Santiago de Chile. *Land Use Policy, 25*(2), 153–160.

Bose, M. (2006). *The Magic of Indian Cricket: Cricket and Society in India*. London: Routledge.

Braut, J. M. (1999). Environmentalism and eurocentrism. *Geographical Review, 89*(3), 391–408.

Carrington, B. (2015). Post/colonial theory and sport. In R. Giulianotti (Ed.), *Routledge Handbook of the Sociology of Sport* (pp. 129–140). London: Routledge.

Carrington, B., & McDonald, I. (2001). Whose game is it anyway? Racism in local league cricket. In B. Carrington & I. McDonald (Eds.), *'Race', Sport and British Society* (pp. 49–69). London: Routledge.

Cashman, R. (1998). The Subcontinent. In B. Stoddart & K. A. Sandiford (Eds.), *The Imperial Game* (pp. 116–134). Manchester, UK: Manchester University Press.

Cole, M. (1982). The beginnings of club cricket. *Wisden Cricket Monthly, 6*(2), 2.

Dussel, E. (1995). *The Invention of the Americas: Eclipse of 'the Other' and the Myth of Modernity*. New York: Continuum.

Fernandes, L. (2004). The politics of forgetting: Class politics, state power and the restructuring of urban space in India. *Urban Studies, 41*(12), 2415–2430.

Fletcher, T. (2011). The making of English cricket cultures: Empire, globalization and (post) colonialism. *Sport in Society, 14*(1), 17–36.

Gould, K. A., & Lewis, L. L. (2017). *Green Gentrification: Urban Sustainability and the Struggle for Environmental Justice*. London: Routledge.

Gregory, D. (2001). Postcolonialism and the production of nature. In N. Castree & B. Braun (Eds.), *Social Nature: Theory, Practice, and Politics* (pp. 84–111). Oxford, UK: Blackwell.

Gregory, D. (2004). *The Colonial Present: Afghanistan, Palestine, Iraq*. Oxford, UK: Blackwell.

Gruneau, R. (2017). *Sport and Modernity*. Cambridge, UK: Polity Press.

Gupta, A. (2009). India and the IPL: Cricket's globalized empire. *The Round Table*, 98(401), 201–211.

Haraway, D. (1989). *Primate Visions: Gender, Race, and Nature in the World of Modern Science*. New York: Routledge.

Kumar, K. (2005). *From Post-Industrial to Post-Modern Society* (second edition). Oxford, UK: Blackwell.

Latour, B. (1993). *We Have Never Been Modern*. Cambridge, MA: Harvard University Press.

Macdonell, A. (1935). *England, Their England*. London: Macmillan.

Majumdar, B. (2007). Nationalist romance to postcolonial sport: Cricket in 2006 India. *Sport in Society*, 10(1), 88–100.

Malcolm, D. (2001). 'It's not cricket': Colonial legacies and contemporary inequalities. *Journal of Historical Sociology*, 14(3), 253–275.

Malcolm, D. (2012). *Globalizing Cricket: Englishness, Empire and Identity*. London: Bloomsbury.

Marqusee, M. (1994). *Anyone but England: An Outsider Looks at English Cricket*. London: Bloomsbury.

Mehta, N., Gemmell, J., & Malcolm, D. (2009). 'Bombay Sport Exchange': Cricket, globalization and the future. *Sport in Society*, 12(4–5), 694–707.

Millington, B., & Wilson, B. (2016). *The Greening of Golf: Sport, Globalization and the Environment*. Manchester, UK: University of Manchester Press.

Mills, J., & Dimeo, P. (2003). 'When gold is fired it shines': Sport, the imagination and the body in colonial and postcolonial India. In J. Bale & M. Cronin (Eds.), *Sport and Postcolonialism* (pp. 107–122). New York: Oxford University Press.

Mitchell, T. (Ed.). (2000). *Questions of Modernity*. Minneapolis, MN: University of Minnesota Press.

Nandy, A. (2000). *The Tao of Cricket: On Games of Destiny and the Destiny of Games*. Oxford, UK: Oxford University Press.

Plumwood, V. (1993). *Feminism and the Mastery of Nature*. London: Routledge.

Pow, C. P. (2009). Neoliberalism and the aestheticization of new middle-class landscapes. *Antipode*, 41(2), 371–390.

Ross, G. (1981). *The Penguin Cricketer's Companion*. London: Penguin.

Sandiford, K. (1984). Victorian cricket technique and industrial technology. *British Journal of Sport History*, 1(3), 272–285.

Silk, M. L. (2004). A tale of two cities: The social production of sterile sporting space. *Journal of Sport and Social Issues*, 28(4), 349–378.

Stoddart, B., & Sandiford, K. A. (Eds.). (1998). *The Imperial Game*. Manchester, UK: Manchester University Press.

Stoddart, M. C. J. (2012). *Making Meaning out of Mountains: The Political Ecology of Skiing*. Vancouver, Canada: University of British Columbia Press.

Stoler, A. L. (1989). Rethinking colonial categories: European communities and the boundaries of rule. *Comparative Studies in Society and History*, 13(1), 134–161.

Waldman, D., Silk, M., & Andrews, D. L. (2017). Cloning colonialism: Residential development, transnational aspiration, and the complexities of postcolonial India. *Geoforum*, 82, 180–188.

Williams, R. (1973). *The Country and the City*. Oxford, UK: Oxford University Press.

Chapter 6

From the IOC Sport and Environment Commission to Sustainability and Legacy Commission

The prospects for the environmental sustainability legacy of Olympic Games host nations under Agenda 2020

John Karamichas

Introduction

The 127th International Olympic Committee (IOC) Session in Monaco (8 and 9 December 2014) reached the unanimous decision to ratify the vision that IOC President Thomas Bach put forward for Agenda 2020. Agenda 2020 entailed changing the IOC Sport and Environment Commission (SEC) to the Sustainability and Legacy Commission (SLC). One of the new principles that underpins Agenda 2020 is to have sustainability permeate all aspects of the planning and organization of an Olympic Games. While the IOC has long sought to align the Games with notions of 'environmentalism', including by naming the environment as the third pillar of Olympism in 1986, Agenda 2020 appears to be making a much stronger claim towards meeting positive outcomes on sustainability in all of its holistic aspects: economic, social, and environmental (IOC, 2014: Recommendation 4).

Before embarking on a discussion of these issues, it is important to highlight that a central aspect of Agenda 2020 is a focus on the environmental sustainability (ES) of the Games. This has also been the focus of earlier studies (see Karamichas, 2012, 2013a, 2013b) that evaluated the post-Olympics capacity for ES by Olympic host nations, particularly through a lens of ecological modernization (EM). Ecological modernization is 'a modernist and technocratic approach to the environment' (Hajer, 1995: 32) and corresponds to weak interpretations of sustainable development (see Christoff, 1996). In many respects, the focus on ES emerged out of the need to identify consistency in the way that the term sustainability is used, and to deal with evident contradictions in its application that have made it suitable for appropriation by disparate suitors, from environmental groups to corporations.

Within the context of the Olympic Games, vague and 'strategically deployable' (see Kirsch, 2010) understandings of sustainability have enabled its use for a variety of means. Particularly in the case of autocratic polities like China, there has been an expectation that hosting the Olympic Games, such as Beijing 2008, would operate in a way akin to Seoul 1988, which led to the enhancement of democratization in South Korea while also improving environmental standards (see Close, Askew, & Xu, 2007; Pound, 2008; Mol, 2010; Mol & Zhang, 2012; McLeod, Pu, & Newman, 2018). However, what has been more common recently is a 'hollowed-out form of sustainable development', such as was seen in London 2012 (Hayes & Horne, 2011: 751), in which the London Games implemented an approach based on EM.

Thus, the focus on ES is in some respects meant to avoid the pitfalls of employing nebulous understandings of sustainable development proffered by contrasting suitors. In other ways, however, Agenda 2020 was developed as a means of counteracting the reputational damage to the IOC after a spate of negative publicity regarding the environmental impact of successive Olympic Games, as well as other issues, including doping, corruption, and human rights abuses (see Horne & Whannel, 2016: 70). In a context where fewer countries are willing to host the Games (Horne & Whannel, 2016; MacAloon, 2016; Geeraert & Gauthier, 2018), the specific negative environmental impacts of the 2014 Sochi Olympics, coupled with human rights issues (Boykoff, 2015; Müller, 2014), clearly stood out as the impetus behind Agenda 2020. Essentially, the very raison d'être of the Olympic Games, and perhaps Olympism in general, were under threat, and Agenda 2020 was presented as an all-encompassing corrective medium.

Given this context, this chapter offers a critical appraisal of the prospects for the ES legacy of the Olympic Games under Agenda 2020 recommendations, and offers an assessment of how the 2012 and 2016 Olympic Games performed in the sustainability realm, and what this means for the future of Olympic hosting, with specific reference to Paris 2024 and Los Angeles 2028. Two overlapping lines of inquiry have organized the endeavour. The first question engages with the identification and use of six EM/ES indicators, and questions the extent to which Olympic Games hosting can lead host nations to enhanced capacity for ES. The identification of these indicators is armoured by an underlying logic that was influenced by Preuss' (2004: 2) call for the need to 'homogenize the calculation methods employed to determine the final balance' when conducting comparative studies of Olympic host countries. The work produced by the team of Burbank, Andranovich, and Heying (2001; Andranovich, Burbank, & Heying, 2001) also influenced the necessary comparative design in answering these questions. The second question pertains to the role of (now mandatory) Olympic Games Impact (OGI) studies in enhancing capacity for ES policies in two host countries, the United Kingdom (2012) and Brazil (2016). Overall, the research draws extensively on official publications by the IOC as well as the various OGI reports. In addition, visits to Sao Paulo, Brazil on two different occasions (2013 and 2015) provided the opportunity to meet and consult with UK Trade and Investment Brazil, during which initiatives to transfer the experience of organizing London 2012 to Brazil were discussed. Insights learned from this fieldwork are also used in the chapter.

The chapter proceeds by offering an overview of the policies developed by succeeding IOC presidents in response to issues of (environmental) concern and crisis points, followed by an account of the post-event ES legacy of summer Olympic Games from Sydney 2000 (the first 'green Olympics') to London 2012, with an emphasis on the *engrenage* dynamic that permeates Olympic Games. *Engrenage* was a centrepiece in the perspective employed by Karamichas (2013b) to assess the post-event capacity for ES policies in four Olympic Games host countries (Sydney 2000/Australia; Athens 2004/Greece; Beijing 2008/China; London 2012/UK). *Engrenage* is a metaphor for the process that sets in motion the coordination of policies by different members to stimulate a specific movement. The chapter then explores the strengths and limitations of Olympic Games Impact reports, with particular attention paid to Rio 2016. These sections set the stage for a detailed account of the changes of Agenda 2020 and its impact on the successful bids of the Paris 2024 and Los Angeles 2028 Olympics. The chapter concludes by suggesting that with the adoption of sustainability in all aspects of the Games, Agenda 2020 appears to operate as a form of enhanced *engrenage*.

From third pillar to OGI and Agenda 2020

If we are to assess the vision of current IOC President Thomas Bach in Agenda 2020, it is worth revisiting the 'visionary plans' of his predecessors by accounting for the environmental factors in them (see Table 6.1). For instance, Jacques Rogge, IOC president from 2001 to 2013, assigned Dick Pound in 2002 to chair a Study Commission with the task of developing a strategic plan for establishing a positive legacy for the Games. Rogge closely followed the steps of his predecessor, Juan Antonio Samaranch (1980–2001), who has subsequently been credited with establishing the environment as the third dimension (or pillar) of Olympism (with sport and culture the other two dimensions), particularly after an environmentally disastrous Albertville Games in 1992 (Cantelon & Letters, 2000; Lesjø, 2000). Indeed, the focus on the environment informed the hosting of Lillehammer 1994, as well as Sydney's 1993 bid to host the 2000 Summer Olympics. However, although promises of 'green Games' have become a hallmark of Olympic hosting, the same environmental challenges continued to appear in the Olympic editions that followed, and played an important role in the formulation of Agenda 2020 (see Table 6.1).

Table 6.1 IOC presidents and environmental sustainability visions

IOC presidents		
Juan Antonio Samaranch (1980–2001)	Jacques Rogge (2001–2013)	Thomas Bach (2013–present)
Vision		
Environment as the third pillar of Olympism: suggested 1986 – official recognition 1994	Positive legacy, impact and sustainable development	Agenda 2020 Sustainability to be included in every single part of an Olympic Games edition
Summer Olympic editions		
Sydney 2000; Athens 2004	Sydney 2000; Athens 2004	Sydney 2000; Athens 2004
Related crisis/development		
– Environmental devastation at Albertville Winter Games – Rio Declaration on Environment and Development – Use of performance-enhancing drugs	– Over-spending – White elephants	– Failures on sustainability goals and environmental harm in Sochi 2014 and Rio 2016 – Use of performance-enhancing drugs by Russian team – Corruption scandals linked to Rio 2016

In terms of reaffirming the environmental orientation of the IOC, the Olympic Games Study Commission, under Pound, produced a report of 117 recommendations in 2003. These recommendations set in motion the centrality of 'legacy' in preparing the bidding applications to host the Games, while focusing on more encompassing definitions of 'sustainable development' instead of the limited focus on 'environmental considerations' (Olympic Games Study Commission, 2003: 5). By legacy, the IOC highlighted the importance of avoiding 'white elephants' by factoring in the post-Olympic use of facilities and sport venues in bid applications.

In response to the increasing environmental focus of the IOC, aspiring Olympic hosts started dedicating whole parts of their applications to the OGI Study. OGI was introduced in 2001 as a central requirement towards substantiating the legacy of an Olympic Games through an 'objective and scientific analysis of the impact' (Economic and Social Research Council [ESRC], 2010: 6). The first OGI study was part of the formal planning requirements for the 2010 Vancouver Olympics and was 'designed to evaluate the Games legacy for the host nation and city against a raft of social, economic, cultural and environmental indicators, hence providing an "evidence base" for measuring the positive societal consequences of the Games for its hosts' (MacRury & Poynter, 2009: 304). While Beijing 2008 voluntarily applied an OGI study, both London 2012 and Rio 2016 were mandated to do so. Overall, it is within this context that the process of accounting for the environmental sustainability of the Games should be understood.

Accounting for post-event capacity for environmental sustainability

The idea of post-event capacity for ES operates on the perception that hosting an Olympic Games is likely to enhance the ability of broader and already existing institutions in the host country to implement ES policies and/or developing capacity for ES (see Weidner, 2002). The totality of the process over the three phases of a mega-event (see Hiller, 2000), from the pre-event phase of bidding to hosting the Games to the post-event phase of ES legacy, was examined by Karamichas (2012, 2013a, 2013b) under the prism of *engrenage*. The underlying rationale is that 'the process of meeting the IOC's environmental standards could both drag with it the host nation's institutional framework and set a precedent that other nations would strive to emulate' (Karamichas, 2012: 156; see also Karamichas, 2013b: 11), in a similar process to the policy considerations in the early days of the European Community.

In assessing post-Olympics ES/EM capacity, we formulated two working hypotheses (see Andersen, 2002; Karamichas, 2012: 152, 2013b: 104) and subsequently tested them by examining six ES/EM indicators based on relevant literature (see Jänicke & Weidner, 1997; Mol & Sonnenfeld, 2000; Weidner, 2002; Buttel, 2003). The ES legacy aspirations of the IOC, the intervening variables of the global economic crisis, and local governmental politics also played a crucial role. Sonnenfeld (2015) reviewed that study and devised a schematic reinterpretation of the findings with a five-point (+2, +1, 0, -1, -2) scale 'representing substantial progress, some progress, no significant change, some regress, substantial regress' (p. 76). The findings show that Sydney and Athens scored negatively overall, while Beijing and London scored the highest.

It is important to note that organizers of the respective Olympics in Sydney, Athens, and Beijing prepared their bids for hosting the Games under different regimes or Manuals for Candidate Cities (MCCs), as compared to the Candidature Procedure and Questionnaires (CPQs) that were employed to formulate the British (London 2012) and Brazilian (Rio 2016) bids. It is also important to reiterate that China was so determined to host the Games after losing to Sydney for the 2000 Games due to its negative human rights and environmental credentials that it proceeded to conduct an OGI study without being obliged to do so.

The data show that Australia and Greece regressed in their capacity for ES/EM in relation to Olympic Hosting. That regress is evident in the Greek case as a direct outcome of the severe economic downturn that the country has been experiencing since 2010. The standard explanation posits that in times of economic crisis, 'bread and butter' issues of concern (jobs and basic income, standard housing, access to basic food and health) take precedence over the post-materialist environmental issues (Maslow, 1970; Inglehart, 1971, 1977, 1990). For both countries, 'the overall conclusion points to a lack of correspondence between the EM aspirations of Olympic Games hosting and the actual capacity for EM' (Karamichas, 2013b: 143). The next section questions whether OGI was also a factor in the sustainability efforts of London 2012 (pre-event and post-event), followed by a discussion of Rio 2016 (pre-event).

Discussing OGI reports: strengths and limitations

The bids behind London 2012 and Rio 2016 had to consider the Olympic Games Impact study which has been a requirement for Olympic Games hosting since 2001. With OGI, the Candidature, Procedure and Questionnaire replaced the Manual for Candidate Cities. Both CPQs and MCCs were operating to inform prospective candidates about what was required in preparing a successful bid. MCCs, with an environment and meteorology section, were used for the first time in 1992 for the Sydney 2000 Olympics, and with minor changes for the Athens 2004 and Beijing 2008 Summer Olympics (see Karamichas, 2013b: 105). Different CPQs guided the bids for London 2012 and Rio 2016. Both reports noted that the scope of the OGI covers a 12-year period (two years prior to the host city election and continuing three years after the Games) and considers three areas of sustainable development – economic, socio-cultural, and environmental – across three impacts zones of country, region, and city (see IOC, 2004; IOC, 2008).

The 2010 OGI study, the pre-Games report, for London 2012, produced only some indications to direct and produce a detailed assessment in the 2015 follow-up study. To give a small indicative illustration, the 2010 OGI report came up with the following findings: 'In both the environmental and economic cases, ... figures reflect the relatively few areas where it is possible to say with confidence that there has been an impact and further an impact that it is due to the Games' (ESRC, 2010: 22).

The conclusions of the 2015 follow-up OGI offered a detailed discussion of 67 indicators and highlighted that London 2012 met its 'sustainability aspirations' with some evident improvements in all three sustainability spheres (environmental, social, and economic). However, the report also cautioned that:

[while] London 2012 has been a catalyst for positive change is not in
doubt … when and where the process ends and what will be the full
magnitude of the effect is not yet known. The story of London 2012
will continue to unfold for a long time to come.

(ESRC, 2015: 185)

Indeed, after reading the final OGI report, the overall impression is
that London 2012 met its sustainability objectives and the study
applied high standards in its execution. This suggests that London 2012
can join Beijing 2008 with a good record of applying sustainability
standards throughout all phases in preparing and delivering the Games.
Still, London 2012 initiated, in part or entirely, several sustainability
instruments that were to be adopted by later Olympic Games editions
and other types of events – such as Event Sustainability Management
Systems, Sustainable Sourcing Code, and Carbon Footprint Methodology, to mention a few (ESRC, 2015: 174–175), but were not taken up
fully. This suggests that:

recognition and uptake [of ES] has been lacking within UK Government and London Government circles and many opportunities to
introduce sustainability management to major events in the capital
and countryside have been missed. … While London 2012 did well to
ensure sustainability sequenced seamlessly from bid to Organising
Committee, there was no such continuity into legacy structures.
A wider learning here is that host cities may actually have more legacy
potential than they realise.

(ESRC, 2015: 176)

Two important issues emerge here. First, London is a global metropolis,
an immensely important nodal point that can direct and influence many
different environmental flows. Yet the sustainability credentials of London
were only either implied or directly indicated when necessary or conveni-
ent, with decisions largely deferred to the European Union. While
'improving air quality in the run-up to London 2012 had been a concern,
the expensive penalties that can arise from not meeting EU standards has
been the dominant driver' (ESRC, 2015: 24). Second, and despite the
limits discussed above, there was something of an uptake of the sustainability standards that were developed and upheld by London 2012, specifically to other 'major events' and areas in London. A six-year report in
2018 gave added confirmation of the highly positive legacy of London
2012 that included regeneration of East London, the clean-up of toxic
areas, the creation of thousands of jobs, and the use of facilities for other
events that generate high profit margins (Rossingh, 2018).

Rio 2016: from boom to bust

Three months after the Rio Olympics, IOC spokesperson Mark Adams called the event 'the most perfect imperfect Games' (Zimbalist, 2017: 2). In that grammatically dubious phrase, there is an element of truth regarding the uncomfortable situation that Rio found itself in post-Games. From the outset, the announcement of Brazil's successful bid in October 2009 came to add to the earlier success of Brazil's bid to host 2014 FIFA World Cup. As then President Luiz Ignacio Lula da Silva proclaimed to cheering crowds, the hosting of two sport mega-events was meant to showcase that '[Brazil's] hour has arrived' (Watts, 2014). Yet one year before the opening of the 2014 FIFA World Cup, in June 2013, Brazil experienced 'the biggest [street protest] in a generation that highlighted dissatisfaction with the dire public services, political corruption, police violence and wasteful spending on stadiums' (Watts, 2014).

A similar protest wave was evident again one year before the 2016 Rio Olympics, with hundreds of thousands gathering in different cities and calling for the president, Dilma Rousseff, to step down. These mobilizations highlighted dissatisfaction with political corruption, but were also linked to the economic downturn the country was experiencing, with the accompanying rising unemployment and inflation rates since 2011 (BBC News, 2015). At the time, the economy was expected to decline around 2%, with inflation nearing 10% and unemployment rising, and anger towards the country's political elite growing (Douglas, 2015). These issues raised serious concerns about the sustainable development claims Brazil made in the bidding application to host the 2016 Games (see also Boykoff, 2017: 180).

The initial Rio 2016 OGI report, published in January 2014, noted that the slowed oil production and the global economic crisis had contributed to the 'worst recession [in Brazil] in more than century, months before the Games commenced' (Boykoff, 2017: 180) and could not have been foreseen (OGI – SAGE/COPPE/UFRJ Research Team, 2014). Indeed, Rio 2016 began its 'Olympic adventure' with some very good odds regarding the potential of meeting the ES commitments in the bidding application and, by extension, stimulating a post-Olympics capacity for ES policies throughout Brazil. This position acquired further credence given the fact that Brazil in general, and Rio de Janeiro in particular, have an intimate connection to global environmental issues. At the 1972 UN *Conference on the Human Environment* in Stockholm, for example, the Brazilian delegation warned of the harms caused by pollution, the combating of which 'only developed countries could afford' (Leonard, 1988: 69; Hogan, 2000: 2), and highlighted the unequal dynamics between the global North and South in regard to industrial development and sustainability. Later, Brazil was also a leader in attempts to compromise these contradictory processes, promoting a 'sustainable development' perspective at the 1992 Rio Summit.

This history informed the sustainable development promises attached to the Rio Olympics. In the bidding application to host the Games, like preceding successful candidates, Rio did not hesitate in making ambitious sustainable development declarations under the general frame of 'Green Games for a Blue Planet', including proclamations by the Brazilian Olympic Committee (BOC) that:

> The Rio 2016 Games in Rio will catalyse the environmental policies and programs of the three levels of government via the Rio 2016's Sustainability Management Plan (SMP). The three pillars of Rio 2016's SMP – planet, people, prosperity – will integrate economic, environmental and social elements into the 'Green Games for a Blue Planet' vision for the Rio Game.
>
> (see BOC, 2009)

Like preceding Olympic Games applicants, Rio also made extensive reference to its existing qualities on the environmental front. On the impact of energy consumption and greenhouse gas emissions, Rio highlighted the importance of the Amazon and Atlantic forests and the efficient use of 'green energy plants and low energy design strategies' for all venues (BOC, 2009). The BOC also made strong claims in regard to the extensive use of 'electrical energy from renewable sources', substantial reduction in the use of CO_2 emissions and the application of 'Brazilian cutting-edge technology initiatives for the use of renewable energy sources during the Games' and a capacity to 'offset the direct emissions of the Games' (BOC, 2009).

It is important to note that Rio also had the assistance of a range of Olympic experts in crafting its sustainability agenda. For instance, Michael Payne, former IOC marketing expert, was recruited as Rio's Senior Strategy Advisor to lead what Clift and Andrews (2012: 219) saw as a 'a cabal of globally peripatetic Olympic bid professionals, whose charge was to create a vision of the Rio 2016 local – within, and through, the bid structure and presentation – that would engage IOC delegates' (see also Pentifallo & VanWynsberghe, 2012: 443; Zimbalist, 2015: 90; Horne & Whannel, 2016: 12). UK Trade and Investment was also set up in the British Consulates in São Paulo and Rio de Janeiro to communicate and share the British experience on sustainability from London 2012. In a meeting at the British Consulate in São Paulo, we heard that the advice was to pursue 'less temporary but "nomadic" structures – e.g. handball arena becoming schools with involvement of UK companies' (UKTI, São Paulo meeting, 22 May 2013).

The pre-event OGI report for Rio 2016 bore many similarities to the London 2012 pre-games OGI report. For instance, there were challenges in accessing data from official records, and in many cases secondary data were used. Compiled a year before the collapse of the economy, and perhaps oblivious to it, the team responsible for the OGI report argued that this 'difficulty will be overcome in following reports …. If access to primary information is not achieved, this could jeopardise the quality of the results and the effectiveness of the focus areas' monitoring' (OGI – SAGE/COPPE/UFRJ Research Team, 2014). We now know that Rio 2016 failed miserably in meeting environmental ambitions, with a significant reputational cost in all counts (see Boykoff, 2017: 189–195). The pre-event standing on ES of Brazil scored positively in two of the ES indicators, environmental concern and CO_2 emissions, without any identified causal connection to the Olympics (see Karamichas, 2015), but by 2015 the economic crisis severely curtailed any positive outcomes on the ES front. Of course, it remains to be seen what the 2019 post-Games OGI report is going to declare. With this in mind, the final section offers a discussion of Agenda 2020 recommendations towards appraising the potential of Olympic Games host nations to acquire and/or maintain capacity for ES policies, and the simultaneous awarding of the 2024 and 2028 Olympiads.

Conclusion: future implications of ES in sports mega-events

To reiterate, Agenda 2020 is in many ways a response that is linked to the following identified negativities:

1   The human rights violations and severe environmental catastrophes of Sochi 2014;
2   The failure of Rio 2016 in all counts, something that had become apparent before 2014;
3   The diminishing appeal of Olympic Games hosting (see Gold & Gold, 2017, Table 2.1, p. 24).

Geeraert and Gauthier (2018: 17) highlight that Agenda 2020, as 'the strategic roadmap for the future of the Olympic Movement, was adopted partly in response to [the aforementioned] pressure'. Forty detailed recommendations make up Agenda 2020. These are described as 'pieces of a jigsaw puzzle, which when put together give us a clear vision of where we are headed and how we can protect the uniqueness of the Games and strengthen Olympic values in society' (IOC, 2014: 1). Furthermore, according to Thomas Bach, Agenda 2020 appears to be very well aligned to the objectives of the 2030 Agenda for Sustainable Development of the United Nations (United Nations, 2015; Boykoff, 2017: 195).

Recommendations 4 and 5 are especially important in advancing an environmental/sustainable focus for Olympic hosting. Recommendation 4 advances a holistic approach to sustainability 'in all aspects of the Olympic Games', including the organization and governance in all phases and legacies of an Olympic Games edition. This is to be supported by Recommendation 5, which includes sustainability within the 'Olympic Movement's daily operations' to promote a sustainability rationale that affects every single component of the IOC's operations, ranging from the simple procurement of goods and services to assisting relevant stakeholders in 'integrating sustainability within their own organisation and operation' (IOC, 2014: 12).

With the increased monitoring that is incorporated in their implementation, these recommendations appear to be an adequate corrective to the inadequacies of the OGI mechanism that allowed for the ES failures of Sochi and Rio. However, the possibility of the host failing to meet sustainability requirements at a late stage in the pre-event phase, months or even a year before the event, has not been factored into Agenda 2020 requirements. The most effective approach to ES would be 'a removal of the Games', but 'the IOC is unlikely to want to cancel or move the Games, because of the enormous financial and reputational costs' (Geeraert & Gauthier, 2018: 26).

The simultaneous awarding of both the 2024 and 2028 Olympics to Paris and Los Angeles respectively in September 2017 (see IOC, 2017) is a further indication of the IOC's ES efforts. The decision was taken after four of the six contenders to host the 2024 Summer Olympics withdrew after expressing concerns about the magnitude of the Games and associated costs. Since the formulation of Agenda 2020 was a response to these concerns, the historic decision to make a simultaneous award of the Games to two bidders on different editions made sense. In the words of Thomas Bach, 'Ensuring the stability of the Olympic Games for 11 years is something extraordinary.' Both cities have also been Olympic Games hosts in the past, with Paris having hosted the Summer Olympics in 1900 and 1924 and Los Angeles in 1932 and 1984. Thus, there is a chance that these events will require less infrastructure development. Indeed, awarding the Games to cities in two 'developed' countries speaks to concerns around the sustainable development promises of the Olympics and notions of development as modernization catalysed through mega-event hosting (see Darnell & Millington, 2016). Could it be that the IOC, after witnessing disappointment from two BRICS (Brazil, Russia, India, China, and South Africa) countries, now prefers the 'safe bet' of some core countries from the global North, instead of countries from the global South or the semi-periphery for Olympic Games hosting? It may also be the case, following the ecological modernization mantra, that modernization can only be good for environmental preservation or in mitigating climatic changes when the socio-political structures are fully modernized and the country is run by an administration that is positively predisposed towards facilitating ES capacity.

Ultimately, regardless of the IOC's motivations, Agenda 2020 foregrounds issues of environmental sustainability for future Olympic hosting. Because Agenda 2020 encompasses sustainability in all three of its component parts (environmental, social, and economic), the best-case scenario would be that Olympic Games hosting would stimulate existing ES capacity in these core countries in the modernization process. Agenda 2020 could cater for multi-city and multi-country Olympic Games editions in future developments. All in all, Agenda 2020 holds potential to operate as a process of enhanced *engrenage* by establishing ES standards within the institutional framework of the IOC, standards to be taken up by host nations through OGI, and which in turn may establish precedents that other nations strive to emulate.

References

Andersen, M. S. (2002) Ecological Modernisation or Subversion? The Effect of Europeanization on Eastern Europe. *American Behavioral Scientist*, 45(9), 1394–1416.

Andranovich, G. D., Burbank, M. J. & Heying, C. H. (2001) Olympic Cities: Lessons Learned from Mega-event Politics. *Journal of Urban Affairs*, 23(2), 113–131.

BBC News. (2015) Brazilian Protesters Call for President Dilma Rousseff's Impeachment, www.bbc.co.uk/news/world-us-canada-33953606, accessed 17 August 2015.

BOC. (2009) *Candidature File for Rio de Janeiro to Host the 2016 Olympic and Paralympic Games*, Volume 1, http://rio2016.com/sites/deafault/files/parceiros/candidature_file_v1.pdf, accessed 10 August 2015.

Boykoff, J. (2015) Sochi 2014: Politics, Activism and Repression. In R. Gruneau & J. Horne (eds) *Mega-events and Globalization* (London: Routledge), pp. 131–148.

Boykoff, J. (2017) Green Games: The Olympics, Sustainability, and Rio 2016. In A. Zimbalist (ed.) *Rio 2016: Olympic Myths, Hard Realities* (Washington, DC: Brookings Institution Press), pp. 179–205.

Burbank, M. J., Andranovich, G. D. & Heying, C. H. (2001) *Olympic Dreams: The Impact of Mega-Events on Local Politics* (Boulder, CO: Lynne Rienner).

Buttel, F. H. (2003) Environmental Sociology and the Explanation of Environmental Reform. *Organization and Environment*, 16(3), 306–344.

Cantelon, H. & Letters, M. (2000) The Making of IOC Environmental Policy as the Third Dimension of the Olympic Movement. *International Review for the Sociology of Sport*, 35(3), 294–308.

Christoff, P. (1996) Ecological Modernisation, Ecological Modernities. *Environmental Politics*, 5(3), 476–500.

Clift, B. C. & Andrews, D. L. (2012) Living Lula's Passion? The Politics of Rio 2016. In H. J. Lenskyj & S. Wagg (eds) *The Palgrave Handbook of Olympic Studies* (Basingstoke: Palgrave), pp. 210–229.

Close, P., Askew, D. & Xu, X. (2007) *The Beijing Olympiad* (London: Routledge).

Darnell, S. & Millington, R. (2016) Modernization, Neoliberalism and Sports Megaevents: Evolving Discourses in Latin America. In R. Gruneau & J. Horne (eds)

Mega-events and Globalization: Capital and Spectacle in a Changing World Order (London: Routledge), pp. 65–80.

Douglas, B. (2015) Brazilian President under Fire as Tens of Thousands Protest in 200 Cities, *The Guardian*, 16 August 2015, www.theguardian.com/world/2015/aug/16/brazil-protests-dilma-rousseff, accessed 17 August 2015.

ESRC. (2010) *Olympic Games Impact Study – London 2012 Pre-Games Report*, https://esrc.ukri.org/files/news-events-and-publications/news/2014/olympic-games-impact-study-london-2012-pre-games-report/, accessed 12 July 2019.

ESRC. (2015) *Olympic Games Impact Study – London 2012 Post-Games Report*, www.uel.ac.uk/schools/ace/research/centre-for-geoinformation-studies/researcharea sprojects/ogispostgameslondon2012, accessed 20 January 2016.

Geeraert, A. & Gauthier, R. (2018) Out-of-control Olympics: Why the IOC is Unable to Ensure an Environmentally Sustainable Olympic Games. *Journal of Environmental Policy & Planning*, 20(1), 16–30. doi: https://doi.org/10.1080/1523908X.2017.1302322.

Gold, J. R. & Gold, M. M. (2017) The Enduring Enterprise: The Summer Olympics, 1896–2012. in J. R. Gold & M. Gold (eds) *Olympic Cities: City Agendas, Planning and the World's Games, 1896–2020*, third edition (London & New York: Routledge), pp. 21–63.

Hajer, M. A. (1995) *The Politics of Environmental Discourse: Ecological Modernisation and the Policy Process* (Oxford, UK: Oxford University Press).

Hayes, G. & Horne, J. (2011) Sustainable Development, Shock and Owe? London 2012 and Civil Society. *Sociology*, 45(5), 749–764.

Hiller, H. H. (2000) Towards an Urban Sociology of Mega-events. *Research in Urban Sociology*, 5, 181–205.

Hogan, D. J. (2000) Socio-demographic Dimensions of Sustainability: Brazilian Perspectives, www.ciesin.columbia.edu/repository/pern/papers/ISARio2000.doc, accessed 11 April 2011.

Horne, J. & Whannel, G. (2016) *Understanding the Olympics*, second edition (London: Routledge).

Inglehart, R. (1971) The Silent Revolution in Europe: Intergenerational Change in Post-industrial Societies. *American Political Science*, 65, 991–1017.

Inglehart, R. (1977) *The Silent Revolution: Changing Values and Political Styles among Western Publics* (Princeton, NJ: Princeton University Press).

Inglehart, R. (1990) *Culture Shift in Advanced Industrial Society* (Princeton, NJ: Princeton University Press).

IOC. (2004) *2012 Candidature Procedure and Questionnaire: Games of the XXX Olympiad in 2012* (Lausanne, Switzerland: International Olympic Committee), https://stillmed.olympic.org/media/Document%20Library/OlympicOrg/Documents/Host-City-Elections/XXX-Olympiad-2012/Candidature-Procedure-and-Questionnaire-for-the-Games-of-the-XXX-Olympiad-2012.pdf, accessed 15 January 2011.

IOC. (2008) *2016 Candidature Procedure and Questionnaire: Games of the XXXI Olympiad* (Lausanne, Switzerland: International Olympic Committee), https://stillmed.olympic.org/media/Document%20Library/OlympicOrg/Documents/Host-City-Elections/XXXI-Olympiad-2016/Candidature-Procedure-and-Questionnaire-for-the-Games-of-the-XXXI-Olympiad-in-2016.pdf, accessed 10 February 2009.

IOC. (2014) *Olympic Agenda 2020: 20 + 20 Recommendations*, https://stillmed.olympic. org/media/Document%20Library/OlympicOrg/Documents/Olympic-Agenda-2020/ Olympic-Agenda-2020-20-20-Recommendations.pdf#_ga=2.91530458.1160689768. 1531716256-1173776011.1527766444, accessed 10 January 2015.

IOC. (2017) How Paris, Los Angeles and the IOC Moulded a 'Win-win-win', www. olympic.org/news/how-paris-los-angeles-and-the-ioc-moulded-a-win-win-win, accessed 10 October 2017.

Jänicke, M. & Weidner, H. (1997) *National Environmental Policies: A Comparative Study of Capacity-building (13 Countries)* (New York: Springer-Verlag).

Karamichas, J. (2012) Olympic Games as an Opportunity for the Ecological Modernisation of the Host Nation: The Cases of Sydney 2000 and Athens 2004. In G. Hayes & J. Karamichas (eds) *Olympic Games, Mega-events and Civil Societies* (Basingstoke, UK: Palgrave Macmillan), pp. 151–171.

Karamichas, J. (2013a) London 2012 and Environmental Sustainability: A Study through the Lens of Environmental Sociology. *Sociological Research Online*, 18(3), 1–6.

Karamichas, J. (2013b) *The Olympic Games and the Environment* (Houndmills, UK: Palgrave Macmillan).

Karamichas, J. (2015) Mobile Policies for Sustainable Development (SD) in Rio 2016: A Preliminary Assessment of SD Capacity in the Pre-event Phase, paper presented at the *More than Just a Game: Mobilities, Infrastructures & Imaginaries of Global Sports Events* conference, Antwerp, Belgium: University of Antwerp, 8–9 October.

Kirsch, S. (2010) Sustainable Mining. *Dialectical Anthropology*, 34(1), 87–93. doi: https://doi.org/10.1007/s.

Leonard, H. J. (1988) *Pollution and the Struggle for the World Product: Multinational Corporations, Environment and International Comparative Advantage* (Cambridge, UK: Cambridge University Press).

Lesjø, J. H. (2000) Lillehammer 1994: Planning, Figurations and the 'Green' Winter Games. *International Review for the Sociology of Sport*, 35(3), 282–293.

MacAloon, J. J. (2016) Agenda 2020 and the Olympic Movement. *Sport in Society*, 19 (6), 767–785. doi: https://doi.org/10.1080/17430437.2015.1119960.

MacRury, I. & Poynter, G. (2009) Olympic Cities and Social Change. In G. Poynter & I. MacRury (eds) *Olympic Cities: 2012 and the Remaking of London* (Farnham, UK: Ashgate), pp. 303–326.

Maslow, A. H. (1970) *Motivation and Personality* (New York: Harper & Row).

McLeod, C. M., Pu, H., & Newman, J. I. (2018) Blue Skies over Beijing: Olympics, Environments, and the People's Republic of China. *Sociology of Sport Journal*, 35 (1), 29–38. doi: https://doi.org/10.1123/ssj.2016-0149.

Mol, A. P. J. (2010) Sustainability as Global Attractor: The Greening of the 2008 Beijing Olympics. *Global Networks*, 10(4), 510–528.

Mol, A. P. J. & Sonnenfeld, D. A. (2000) Ecological Modernisation around the World: An Introduction. In A. P. J. Mol & D. A. Sonnenfeld (eds) *Ecological Modernisation around the World: Perspectives and Critical Debates* (London: Frank Cass), pp. 3–16.

Mol, A. P. J. & Zhang, L. (2012) Sustainability as Global Norm: The Greening of Mega-Events in China. In G. Hayes & J. Karamichas (eds) *Olympic Games, Mega-events and Civil Societies* (Basingstoke, UK: Palgrave Macmillan), pp. 126–150.

Müller, M. (2014) (Im-)mobile Policies: Why Sustainability Went Wrong in the 2014 Olympics in Sochi. *European Urban and Regional Studies*, 22(2), 191–209. doi: https://doi.org/10.1177/0969776414523801.

OGI – SAGE/COPPE/UFRJ Research Team. (2014) *Olympic Games Impact Study – Rio 2016. Initial Report to Measure the Impacts and the Legacy of the Rio 2016 Games*, www.kennisbanksportenbewegen.nl/?file=8014&m=1500889249&action=file. download, accessed 20 February 2014.

Olympic Games Study Commission. (2003) *Report to the 115th IOC Session by Richard W. Pound*, https://stillmed.olympic.org/Documents/Reports/EN/en_report_725. pdf, accessed 15 January 2012.

Pentifallo, C. & VanWynsberghe, R. (2012) Blame It on Rio: Isomorphism, Environmental Protection and Sustainability in the Olympic Movement. *International Journal of Sport Policy and Politics*, 4(3), 427–446. doi: http://dx.doi.org/10.1080/19406940.2012.694115.

Pound, R. (2008) Olympian Changes: Seoul and Beijing. In M. Worden (ed.) *China's Great Leap: The Beijing Games and Olympian Human Rights Challenges* (New York: Seven Stories Press), pp. 85–97.

Preuss, H. (2004) *The Economics of Staging the Olympics: A Comparison of the Games 1972–2008* (Cheltenham, UK: Edward Elgar).

Rossingh, D. (2018) How the London Olympics Still Generate $176 Million Six Years on from Opening Ceremony, www.forbes.com/sites/daniellerossingh/2018/07/29/how-the-london-olympics-still-generate-176-million-six-years-on-from-open ing-ceremony/#4b6fa5111113, accessed 30 July 2018.

Sonnenfeld, D. A. (2015) Review of the Olympic Games and the Environment. *Contemporary Sociology*, 44(1), 75–77. doi: https://doi.org/10.1177/0094306114562201z.

United Nations. (2015) *Transforming Our World: The 2030 Agenda for Sustainable Development*, New York: United Nations, https://sustainabledevelopment.un.org/content/documents/21252030%20Agenda%20for%20Sustainable%20Develop ment%20web.pdf, accessed 29 June 2019.

Watts, J. (2014) Voices of Brazil: 'Our Hour Has Arrived', *The Observer*, 26 January, www.theguardian.com/world/2014/jan/26/voices-of-brazil-our-hour-has-arrived, accessed 15 March 2014.

Weidner, H. (2002) Capacity Building for Ecological Modernization: Lessons from Cross-national Research. *American Behavioral Scientist*, 45(9), 1340–1368.

Zimbalist, A. (2015) *Circus Maximus: The Economic Gamble Behind Hosting the Olympics and World Cup* (Washington, DC: Brookings Institution Press).

Zimbalist, A. (2017) Introduction: 'Welcome to Hell'. In A. Zimbalist (ed.) *Rio 2016: Olympic Myths, Hard Realities* (Washington, DC: Brookings Institution Press), pp. 1–11.

Chapter 7

Ecological modernization in 2018 PyeongChang Winter Games

The elitist and unjust environmental performance

Kyoung-yim Kim

Introduction

Since the International Olympic Committee (IOC) included the "environment" as the third pillar of the Olympic Movement in 1994, the Olympic Games have become a novel site in which the international policy agenda of environmental sustainability has engaged with national environmental policies through the hosting of the event. Such was the case when South Korea (hereafter, Korea) hosted the 2018 PyeongChang Winter Olympics. Since the beginning of 2000, Korea had ranked sixth highest in greenhouse gas (GHG) emissions, following China, the USA, India, Japan, and Germany. Far worse, Korea has recorded the world's highest increased rate of carbon emissions over 20 years: from 1990 to 2010, emissions increased 146%, far higher than the next highest offenders, Chile (125%), Turkey (109%), and Israel (103%) (GHG Statistics of Korea, 2015). Thus, from the bidding stage, the Korean Government sought to align its bid with issues of environmental awareness by framing the Games as a pathway to sustainability for PyeongChang and Korea more broadly.

When the 2018 Winter Games were awarded to PyeongChang, a Sustainability Team was organized in 2013, and the team, in turn, recommended and adopted ecological modernization strategies for sustainable development of and through the Games. The sustainable development strategies were organized under five central themes, and two among the five themes—Low Carbon Green Olympics and Stewardship of Nature—were environmental concerns. This chapter pays attention to the two environmental themes in the Korean Government's sustainability goals and discusses how those were implemented through the Games. Further, it investigates the current intersection of Korea's promise of environmentally sustainable development through Olympic hosting and the international policy agenda of climate crisis, while also discussing the challenges and tensions that remain within the local context.

Olympic-led environmental sustainability in Korea

The focus on environmental sustainability by the Korean Government and the PyeongChang Organizing Committee for the 2018 Olympic and Paralympic Winter Games (POCOG) was critical in winning the bid to host the Games (Preuss, 2013). The environmental discourse throughout the bidding process drew upon ideas, rhetoric, and strategies of ecological modernization (Kim & Chung, 2018), as the IOC recommended adherence to the standards and principles of eco-modernist environmental sustainability (see Gaffney, 2013; Karamichas, 2013; Wilson & Millington, 2013). Ecological modernization (EM) views environmental degradation caused by capitalism as an inevitable but treatable part of the development process (e.g., Buttel, 2000; Mol & Sonnenfeld, 2000; Mol & Spaargaren, 2002). Thus, EM simultaneously promotes a commitment to economic growth as well as sustainable improvements in environmental practices with the help of science and technology.

The Korean government and POCOG's EM-centered commitments to environmental sustainability were articulated in their two environmental goals—Low Carbon Green Olympics and Stewardship of the Nature. In 2009, then-President Lee Myung-bak declared that Korea would voluntarily reduce GHG emissions by 30% before 2020. POCOG also announced efforts to assist the government by minimizing the carbon footprint of the Games, so the country could meet its national and international targets for the reduction of GHGs. The government—first under Lee, then Park—seized upon a Green Growth initiative based on the idea that ecologically friendly development would enhance Korea's international image, protect the environment, and boost its economy simultaneously (Green Growth Committee, 2009; Ha & Yoon, 2010). Following the PyeongChang Games' first theme of environmental sustainability, the Korean Government named the green growth initiative as a Low-Carbon Green Growth Project (*Jeo Tanso Noxsak Seongjang Jeong Chak*) and introduced various financial incentives for corporations to improve their environmental policies (e.g., an emissions trading scheme).

Low-carbon green Olympics

Reducing carbon emissions was a top priority in Korea's approach to environmental sustainability at the PyeongChang Olympic Games. In order to realize the low-carbon Olympics, the POCOG established seven tasks: (1) reduce and offset GHG emissions, (2) build green transportation, (3) design and construct sustainable venues, (4) use renewable energy, (5) practice green procurement, (6) foster environmental awareness, and (7) communicate with stakeholders (POCOG, 2017, pp. 50–65). Each are discussed here.

Reduce and offset GHG emissions

In its pre-Games' sustainability report, POCOG pledged that the PyeongChang Olympics would be the first Winter Games to go beyond "zero emissions" and aim for O_2 Plus[1] by reducing and offsetting GHG emissions (equivalent to 1,596,000 tons of CO_2). POCOG set about realizing the O_2 Plus vision by establishing a Certified Emission Reduction (CER) program that promotes public donations for carbon credits, afforestation, and creating carbon offset funds. By December 2017, POCOG had received 93.1 million tons of carbon credits from seven private companies and public institutions, including Solvay Korea[2] and KangWon Wind Power, as well as the Korea District Heating Corporation. POCOG also opened the PyeongChang Sustainability Website, which provided real-time information on the state of the environment and greenhouse gas emissions. The website—a first for an Olympics—provided detailed information on GHG emissions, air quality, indoor air quality, and water quality. The Environment & Greenhouse Gas Information System on the website converted energy usage in real time to GHG emissions and confirmed carbon emissions and reductions in real time throughout the Olympic competition.

Build green transportation

For the Olympic Games, Korea built low-carbon transport systems such as the high-speed railway between WonJu and GangNeung, and introduced eco-friendly vehicles (e.g., electric and hydrogen vehicles) throughout the venues along with emissions monitoring systems. The high-speed railway that connected Incheon International Airport with PyeongChang and GangNeung via Seoul was meant to improve visitors' accessibility to Olympic venues as well reduce the carbon footprint by reducing the number of gasoline vehicles on the road during the Games. POCOG estimated that if 420,000 visitors chose to use the WonJu-GangNeung express railroad instead of personal vehicles, it would reduce GHG emissions by 6,654 tons (POCOG, 2017, p. 56). POCOG also used eco-friendly vehicles and set up charging stations with the help of the Korea Electric Power Corporation (KEPCO). KEPCO provided an additional 150 electric vehicles, 24 charging stations, and 15 hydrogen-powered vehicles, maintained in collaboration with public and private actors such as the Ministry of Trade, Industry and Energy, GangWon Province, and Hyundai Motor Company. According to POCOG, these transportation systems will reduce carbon dioxide emissions by 404,000 tons, which could offset 133,500 tons of carbon dioxide, or 84% of the projected emissions from the Games. In addition, POCOG introduced environmentally friendly methods of snow removal instead of using sodium chloride, which often causes ecological damage.

Design and construct sustainable venues

To prepare for the Games, POCOG constructed six new venues and refurbished six others in three regions of the province: PyeongChang and JeongSeon counties and GangNeung city. For those new facilities, POCOG encouraged builders to reduce carbon emissions from heavy equipment and use green products in construction. The six newly built stadiums—the Olympic Sliding Center, GangNeung Oval, GangNeung Ice Arena, GangNeung Hockey Center, KwanDong Hockey Center, and the JeongSeon Alpine Center—received Green Building Certification (GBC) and Energy Efficiency Certification awards. The GBC is based on the Green Standard for Energy and Environmental Design (G-SEED). This system—a rating tool for buildings that consume less energy and reduce pollution—was developed by the Korean Ministry of Land, Infrastructure, and Transport and the Ministry of Environment. The two certifications which are awarded by the Korean Government promote facilities that generate and use renewable energy (e.g., sunlight, solar, and geothermal heat), save energy (e.g., with green roofs, insulation, and airtight doors and windows), conserve water (with cisterns to collect rainfall and water circulation systems to heat and cool buildings), and preserve ecological wetlands and permeable blocks.

Use renewable energy

The POCOG mandate to use renewable energy sources meant heavy investment in geothermal, solar, wind, hydroelectricity, and hydrogen energy. The six newly constructed competition venues accommodated solar and geothermal generation facilities. Solar power was used to generate electricity, and geothermal energy heated the water supply. Wind turbines were also installed in the new stadiums. About 12% (around 4,564 kW per day) of the stadiums' energy needs were supplied through green-generated energy[3] (POCOG, 2017, p. 61). POCOG also extended its promises to include using renewable energy to power the three host cities (PyeongChang, GangNeung, and JeongSeon) during the Olympic Games (altogether, an estimated 243 MW per day). Specifically, the three counties and city produced 143.4 MW of power from 73 wind turbines, and 30 new power plants were added to the plant to secure a total of 59.65 MW. A total of 48 generators were planned to be completed before the opening ceremony of the Olympic Games. If the corresponding 44.7 MW had been secured, the power consumption would have exceeded 243 MW.

Practice green procurement

In an effort to promote sustainable consumption, POCOG followed the *Green Procurement Guideline* and the *Guideline for Selection of Eco-friendly Sponsors* when it selected corporate partners and sponsors. Based on the guidelines, POCOG encouraged related companies to enter into a voluntary agreement which prompts the use of low-carbon materials in venue construction or to use construction equipment generating low carbon. Further, in sponsor selection POCOG prioritized ISO14064-1-certified[4] companies and other companies that manufactured and supplied green products.

Foster environmental awareness

POCOG established several programs that raised environmental awareness. The committee produced and distributed (upon request, since February 2016) approximately 20,000 leaflets to elementary schools, with educational materials on environmentalism and the Winter Games (*Environmental Trip to PyeongChang Winter Olympics*).

Communicate with stakeholders

POCOG communicated with various stakeholders to achieve its green goals: the International Olympic Committee/International Paralympic Committee, National Olympic Committee/National Paralympic Committee, National Assembly/Government, local communities, sponsors, and partners, suppliers, and spectators and citizens. For the public–private communication program, the Committee on Environment and a Green Management Council were established and operated as public agencies to communicate government environmental goals. In return, private stakeholders among the Olympic sponsors (e.g., LG, POSCO, Daelim Construction & Petrochemical Company, and Samsung) promised to help to achieve the government's environmental goals in the Olympics, their tactics including an intelligent traffic system, energy saving technologies, smart highways, renewable energy, and socially engineered behavioral changes.

Stewardship of nature

GangWon Province, where the three Olympic cities are located, has been largely defined by its preserved natural resources and cultural heritage. The government and POCOG recognized the environmental significance of the area and the potential impact of an Olympic Games. Mountains and forests account for 84% of the total land area in the province, hence the region is often referred to as the "lungs of Korea." Thus, stewardship of nature was the other priority of POCOG's environmental sustainability program, which spoke of "ecological restoration of the regions and maintaining biodiversity ... management of the atmosphere, noise, waste, and water quality ..." (POCOG, 2017, p. 67). To achieve this environmental goal, POCOG established five action plans to: (1) conserve biodiversity, (2) restore ecology, (3) manage air quality and noise, (4) manage waste, and (5) manage water quality and sewage treatment (POCOG, 2017, pp. 66–79).

Biodiversity

The Korean Government adopted the United Nations Convention on Biological Diversity (CBD)[5] in May 2014, and as a signatory and ratified member, established the Third National Biodiversity Strategy and Action Plan to fulfill its domestic obligations. The Korean Government carried out this domestic obligation by hosting the *12th Conference on Biological Diversity* in September 2014 in the Olympic city, PyeongChang, and the meeting adopted the PyeongChang Road Map, which addresses ways to achieve biodiversity through technology cooperation, funding, and strengthening the capacity of developing countries (UN CBD, 2014). In Korea, the issue of biodiversity was brought to the forefront during construction of the Alpine Ski venue on Mt. Gariwang. Local nongovernmental organizations (NGOs) raised serious questions about the economic and environmental costs of constructing such a venue on the mountain. The NGOs held a series of anti-development press conferences, public debates, and protests. They also launched a petition against development that received strong support internationally. Due to the concerns that were raised, POCOG and the local government of GangWon Province awarded a contract to a private company to perform the environmental impact assessment on the downhill venue. Instead of building separate facilities for the men's and women's downhill courses, the PyeongChang Winter Olympics combined the two courses—a first in Olympic history. The courses were combined to avoid seven major vegetation habitats that contain protected species. The starting point in the venue was lowered from the peak, Jung-bong, to Ha-bong on Mt. Gariwang, which reduced deforestation by 25 hectares.

Ecology

Along with conserving biodiversity, POCOG committed to preservation of the local ecosystem, and when that was not possible, environmental restoration and repopulation. This included securing endangered wildlife, topsoil preservation by creating alternative forests and landscapes near the venues, restoring streams that had been diverted and creating ecological exploration trails, and creating alternative protection areas. Specifically, POCOG restored 174 hectares of forests beyond the area of 156 hectares promised in its bid commitment. From 2015 to 2016, 9.3 billion South Korean won (hereafter, won) were invested in forest restoration and creation of scenic and replacement forests. When GangWon Province constructed a venue, it collected the topsoil (10,886 m^2) during the construction period, and the remaining

5969 m^2 were stored underground to be used for successful restoration of the ecosystem. Endangered animals (e.g., long-horned beetles, Manchurian trout, long-tailed gorals, and Korean rat snakes) were reintroduced to the areas. Further, prior to the construction of the JeongSeon Alpine Center on Mt. Gariwang, 272 trees (including Mongolian oaks, caster aralias, and yews) and rare herbaceous plants (including alpine leek, white woodland peony, wake-robin, and Korean anemone) were removed and transplanted to 54 sites.

Air and noise

To manage air quality and reduce noise pollution inside and outside the host region, POCOG designated Low Emission Areas around the venues, and Green Buffer Zones within the Olympic villages. The venues were smoke-free zones, to ensure clean air.

Waste

Minimizing waste was managed by providing recycling bins. For collecting recyclable waste, POCOG signed an agreement with the Korea Circulation Resource Distribution Support Center. The government also let a contract to a third-party service provider, and the provider collected the waste through the Albaro System,[6] an integrated information system for managing the generation, transportation, and disposal of construction waste. The program issues certificates to each disposer, carrier, and processor. Recyclable garbage was collected from stadiums and athletic villages and taken to the distribution center by the service providers. There, the recyclables were separated into six categories: paper, glass, cans, PET bottles, plastic, and Styrofoam. The Ministry of Environment Korea monitored the transport and treatment processes on a real-time basis using the electronic waste management system.

Water and sewage

GangWon Province and municipalities ran water quality monitoring and reuse systems in the venues. In addition, POCOG consolidated sewage treatment facilities and management systems in the region. To secure the necessary clean water, water reuse systems with a total storage capacity of 2,377 tons were installed in the four venues (GangNeung Oval, GangNeung Ice Arena, GangNeung Hockey Center, and KwanDong Hockey Center). After the water was treated, water quality was tested on a regular basis, then it was used for irrigation around the buildings. Further, water saving components were installed in faucets, toilets, and showers to reduce water use. POCOG also collaborated with Coca-Cola Korea and the World Wide Fund for Nature to launch the Integrated Water Resources Management Project.

Wastewater emissions, especially snow removal chemicals and deicers, were a major concern. POCOG, along with the Ministry of Environment Korea, purchased eco-friendly snow removal chemicals with low or zero-chloride contents and halted the supply of calcium chloride and salt to the market.

Tensions between global environmental concerns and local practices

With its emphasis on environmental sustainability, the Games have resulted in advancing and modernizing Korean environmental policy and action plans with an EM ethos. The Korean Government's greening efforts for a low-carbon Games and stewardship of nature were similar to the IOC's environmental sustainability indicators, yet the Korean Government's prioritization and implementation tasks were different. In March 2005, the IOC and its Sport and Environment Commission updated eight environmental issues and sustainability indicators in the Olympics that are based on the agenda of the United Nations Environment Programme: biodiversity conservation, ecosystems protection, land use and landscape, pollution, resource and waste management, health and safety, nuisances, and safeguarding of cultural heritage. Korea's firm engagement with the international sustainability agenda and its consideration of local environmental conditions transformed the focus and practices in these areas. However, in the post-Olympic period the harmful environmental effects of hosting have come to light and indicate that the government and POCOG maintained poor records about sustainability efforts, that there were strong differences of opinion among key stakeholders, and negative consequences on the environment of building for the Games.

Business-as-usual in sports and its impact on the local environment

Gaps between what was said and done by the government and POCOG appeared soon after the Games started. The most visible and serious report was from the alpine ski slopes on Mt. Gariwang. Prior to the Olympic Games, NGOs criticized the potentially massive deforestation, soil erosion, loss of soil stability, and destruction and modification of natural ecosystems. In response, POCOG strengthened its restoration plan. However, *Yonhap News* (2018) reported that the restoration plan was unrealistic and already failing when the Games began. On February 21, 2018, one news report described it this way:

> The soil layer, which was promised to use for restoration, was buried in the slope as it was hard to use for restoration. The existing topsoil that was left to be used for restoration was stacked in such a way that the soil's vitality could not be maintained. GangWon Province and

construction companies said they only followed the guidelines of the Ministry of Environment and the Ministry of Forestry's consultation regarding the management of transplanting trees and topsoil. The implementation and monitoring of the environmental impact assessment during the venue construction was totally insufficient. The two ministries did not manage the construction carefully .... They did not understand the ecological damage [caused by the construction of the ski slopes].

(*Yonhap News*, 2018)

Along with the topsoil loss and poor management of soil layers for restoration, the news report described a series of damages in Mt. Gariwang that includes the withering of hardwoods and conifers caused by disturbances to the ecosystem (e.g., the disturbance of underground water flow) and unnecessary logging for convenience of construction companies. Thus, while local stakeholders were key contributors to such environmental harm, Olympic standards also contributed to POCOG's failure to meet its sustainability promises. Mt. Gariwang was chosen as the location for the Alpine Speed Venue because it was the only mountain in the area that met the international requirements for Olympic competition (height, length, elevation, and fall). A group of experts proposed alternatives due to Mt. Gariwang's unique and rich ecosystems and the difficulties restoration might pose. The international authorities rejected the alternatives. Green Korea United, a Korean environmental NGO, began criticizing the decision when the venue was in the planning stages:

Based on the Olympic standards for alpine ski competitions, a venue with an elevation of 800 to 1,100 meters is required. To meet this standard [elevation that is directly connected to height of a mountain], that means a mountain of 2,000 meters or higher. Even Mt. Gariwang, at the height of 1,561 meters, did not measure up to the slopes at the 18th Nagano Winter Olympic Games to the 23rd Sochi: Nagano's downhill slope was at 2,696 meters in height; 2002 Salt Lake City's was 2,917 meters; and 2006 Torino's was 2,823 meters. Eventually, excessive deforestation is required in order to meet the 800-meter altitude difference [by building ground up high with an embankment] claimed by the International Ski Federation.

(Green Korea United, 2015b, p. 34)

The mountainous terrain in Korea is distinct from other Asian, North American, and European countries where the international standards of alpine ski competitions were set. Guided by the International Ski Federation, POCOG managed to build a stadium that met these standards, although local NGOs called it an environmental crime. Thus, while sports communities are making changes in response to climate change (e.g., corporate social responsibility projects), international sports institutions like the IOC are subscribing to a "business as usual" approach that promotes one-size-fits-all standards which can ultimately cause environmental damage, as they did in PyeongChang.

These tensions between the sustainability promises of the IOC and their practical application have been discussed in the academic literature on the topic. Hosting Olympic events stimulates economic growth, improves transportation infrastructure and cultural facilities, and enhances global recognition and prestige, according to Essex and Chalkley (2004). While those intangibles are well-known Olympic legacies among many scholars, others such as Lenskyj (2008) and Boykoff (2014) have argued that the benefits of hosting an Olympic Games are not economically sustainable. Similarly, Giulianotti, Armstrong, Hales, and Hobbs (2015, p. 103) have called the Olympic Games a neo-liberal festival of capitalism "organized to advance private, commercial, and free-market interests ... through vast public spending on facilities, infrastructure, and wider regeneration policies in urban spaces." Rather than benefitting the local economies of host countries, the Games benefit only the IOC and large transnational sponsors, thereby escalating global and local inequalities and leaving the environmental damage for local residents to clean up.

In this regard, the economic outcomes of Olympic hosting are often prioritized over their environmental impacts. For example, the promise to promote the PyeongChang area as the Asian hub of winter sports is a typical economic model of development through sport mega-events. The urban planning and economic revitalization surrounding the Olympic project increased the private sector's capital investment in the province. The emphasis on economic growth and development through sport, however, overshadowed ecologically sound development. In the process of building the winter sports belt, trees, flora, and fauna were transplanted along with residents, who made up the local labor force for agricultural, forestry, mining, and manufacturing concerns. The environmental consequences took a back seat to the promise of building a lasting sport-tourism destination, and as previous Olympic Games have demonstrated, local residents often suffered the after-effects, such as pollution, dangerous chemicals, and damaged land. This echoes Mangan's (2008) review of post-Olympic legacies that are unfulfilled, Terret's (2008) case study of the 1992 Albertville Games, and Beder's (1999) criticism of the environmental destruction of the local community that emerged during and right after Australia's Olympic bid.

Environmental overconsumption and the unjust burden on locals

During the Olympic Games, resources are often overconsumed and products that cause environmental damage are introduced into the local ecosystem. In

Korea, reported overconsumption included water (from rivers, lakes, and the ground), chemical contaminants (icing, deicing chemicals, and explosive and toxic ammonia/methane [CH_4] used on bobsled and luge tracks), land, energy, and more (Green Korea United, 2015a). Water exhaustion is a particular threat, not only to the surrounding ecology, but also to the local community and industry in PyeongChang and beyond. Local environmental groups reported that an excessive amount of water was used to make ice and artificial snow in the various venues during preparations for the Games. When groundwater is overconsumed, the problems may extend beyond an immediate shortage of water for drinking or agricultural needs. Rocks may sink, lowering the water table even further and making it difficult for agricultural concerns to obtain the water they need for crops. NGOs urged the municipalities to prepare alternatives for the excessive water use during the Olympic Games.

Industrial deicing chemicals (e.g., calcium chloride, industrial salt) that cause environmental damage were also overconsumed. Even though POCOG and the province promised and put forth efforts to use eco-friendly deicers, the unusually low temperature and heavy snow in the region coupled with the increased cost and inefficiency of environmentally safe deicers led to the use of the more dangerous chemicals such as industrial deicers. By way of comparison, the price of industrial calcium chloride is 180 won per kilogram, but an eco-friendly deicer costs 350 won per kilogram. Further, despite the higher cost, the performance was poor or slow, therefore local governments reported that they used industrial deicers at the risk of environmental problems (*Yonhap News*, 2016). The unaffordable resources and overconsumed chemicals were thus contributing factors to environmental damage.

Preuss (2013) contended that the financial shortcomings and the event organizer's lack of serious interest in the environment impeded the production of a green Games. However, this investigation simplifies environmental performance in sport as a matter of budget and awareness, and ignores the structural problems that arise between the international sports organization and the host countries. Korea was not an isolated case. Similar experiences have emerged from previous Winter Games. For example, Timsheva (2001) investigated the environmental legacy of the 1992 Albertville Winter Games in France and reported exhaust pollution, deforestation, and erosion of the alpine mountains following construction of the biathlon and ski racing tracks. Findling and Pelle (1996) also found the 1992 Games left a "legacy of pollution and environmental injury" because of the poisonous ammonia use in the bobsled and luge venue. Similarly, massive deforestation, pollution, destruction of the ecosystem, and other environmental problems were reported following the 1998 Nagano Winter Games (Ezawa, 2015). These recurring problems indicate that such issues are not contextually confined to a host nation, but rather endemic to Olympic hosting itself.

Although sufficient green and renewable energy was produced for the Olympic Games, according to POCOG, it is unclear if the clean energy resources continue to be generated for the local communities. Kim (2018), an energy welfare expert, contends that the supply of heating and cooking gas used by locals is indicative of the energy welfare of the rural and mountainous regions of Korea. He finds that energy welfare in the mountainous PyeongChang area is poorer than in other parts of Korea; however, all of the newly built energy infrastructure in the province for the Olympic Games would not enhance local energy welfare without further costs. The infrastructure was for large commercial venues, and the conversion from industrial to household energy demands time and an immense budget. Kim (2018) argues that although the post-Olympic infrastructure will be maintained partly by the local taxpayers, it is unclear whether the residents will benefit from the energy infrastructure. Thus, the overconsumption and the lack of sustainable plans have unjustly imposed long-term burdens—environmental, economic, and social—on residents.

Economic calculation of nature and market actors' self-regulation

The dominant global environmental discourse prioritizes reducing GHG emissions as an urgent action in response to climate change. POCOG's claim of achieving an O_2 Plus Games through the carbon emission trading scheme was an economic calculation to solve environmental problems. Scholars argue that carbon offsets are a new form of commodification and regulation of nature through environmental governance under neoliberalism (e.g., Bridge, 2002; Gibbs, 2006; Bumps & Liverman, 2008). In particular, Bumps and Liverman (2008) identify the ways in which carbon offsets create new commodities and markets that connect the global North and South, corporations and consumers, environmental groups and transnational institutions, and science and markets. By emphasizing the inequalities of environmental economic geography, these scholars criticize carbon offsets as:

> sloppy definitions of additionality and development benefits, for neocolonial practices of unequal exchange and the dispossession of rights in selling cheap credits to the North obtained from projects in the South, and for the lack of transparency and participatory governance.
> (Bumps & Liverman, 2008, p. 148)

Through this lens, the PyeongChang Games' achievement of "zero" carbon emissions through carbon trading and donations by corporate partners is ambiguous and misleading.

Further, the economic calculation of nature and the environment around the Olympic Games is problematic, especially when reflexivity in environmental performance relies on self-regulation among the market actors. The technocratic aspects of ecological modernization discourse refer to the translation of social and moral ecological issues into marketplace issues. Hajer (1996) coined the term "technicization" to refer to the eco-modernists' use of scientific discourse to solve environmental problems through innovations and investment. Ecological modernization engages in various initiatives to address environmental problems while boosting the environmental industry sector (OECD, 2007, p. 43). The Korean Government and POCOG's eco-industrial innovations, with the support of green science and technology, flourished during the Games, especially in transportation, venue construction, and energy generation. With environmental promises and plans to optimize these innovations for the Winter Olympics, the government established amicable relationships with market players such as LG, Samsung, POSCO, DaeLim Construction & Petrochemical Company, and DaeJung Precision Co., Ltd., with some obtaining patents for green technologies invented for the Games. For example, the Korean Government signed a turnkey contract[7] with DaeLim Construction & Petrochemical Company to construct the Alpensia Sliding Center. The company then developed cutting-edge and eco-friendly (reduced construction periods radically, and thus contributed to less GHG emissions) sledding tracks and earned international patents. The turnkey contract had the merit of unifying the responsibilities and taking advantage of the new technologies possessed by the construction company. However, it included a fatal disadvantage—the contractor could make arbitrary decisions, as it was responsible for evaluating the environmental impact and gathering opinions from the residents. Indeed, the company was caught engaging in large-scale woodcutting in the preserved green lands for the convenience of construction; as a result, the PyeongChang county office sued it (Sisaweek, 2014). While technocrats within the private sector have implemented the scientific management of wildlife and natural resources to achieve optimal economic yields, it nevertheless undermines the local environment.

The case of PyeongChang demonstrates that ecologically ignorant economic interests hinder the diffusion of environmental innovations. Ecological modernization critics (e.g., York & Rosa, 2003) also question whether green consumption through technological advances alone can achieve resource conservation and better environmental protection, particularly if left to businesses to self-regulate. As York and Rosa have pointed out, it is difficult to expect self-reflexivity on environmental performance when feasibility is emphasized within the context of affordable (cheap) technology and maximum profit-making. Similarly, this challenge occurred in PyeongChang's case when reflexivity in environmental performance was left up to the business sector. As discussed earlier, DaeLim

Company, an official business partner with POCOG, destroyed preserved lands and forests for the sake of convenience and reduced costs in venue construction (*Sisaweek*, 2014). The capitalist logic of choice can be easily contrasted to the reflexive act of choosing a more environmentally friendly practice.

An esthetic-oriented, elitist model of sustainability

The Korean Government's environmental policies and institutional reforms before and during the Olympic Games are what Karamichas (2013, p. 133) termed the "eco-modernist institutional amendment." The ecological modernization paradigms in PyeongChang's Olympic Games advanced environmental reform, promoted the use of green production and recycled material, and improved policing of environmental performance. The eco-modernist institutional amendments and bureaucratic restructuring followed the vitalization plan for the local economy. That plan was problematic, as local economic development often took precedence over environmental protection. This led to the implementation of a weak model of sustainability. While the strong model of sustainability insists that we should conserve and enhance our natural capital stocks and live on the income generated by them, the weak model of sustainability holds the position that we can lose natural capital if we substitute the equivalent "human capital" (e.g., scientific invention, technological innovation of resources and others) (Agyeman, 2013).

PyeongChang's weak model of environmental sustainability appears to be largely rhetorical and esthetic. Besides the emphasis on economic development, POCOG's environmental promises to conserve biodiversity and restore ecology proved to be nothing but talk. The total activities during the Olympic preparation period, especially from 2015 to 2016, consisted of afforestation (278 hectares), street tree planting (63 hectares), and scenic forestation (629 hectares) (POCOG, 2017, p. 71), which focused on esthetics rather than restoration of the damaged ecology. This esthetic restoration and lack of long-term, reflexive afforestation plans lead to weak environmental sustainability. As for the esthetic environmental performance, Lubbers (2002) contends that the goal of corporate-driven environmental works prioritizes the appearance of environmental-friendliness, which he calls "green capitalism." Each element of green science and eco-friendly technologies (e.g., carbon offsets, G-SEED-certified stadiums, and intelligent energy systems) is not only dominated and organized by corporations and capital mathematically, but each is also repeating the old patterns of urban-centered interests that marginalize locals and the needs of their communities.

POCOG's ecological modernization paradigms also facilitate the top-down style of environmental policy and governance. Rajkobal (2014)

﹍escribes the policy approach of ecological modernization as expert-oriented and science-based, and one that has become an effective source of social control in modern societies; both governments and other authorities use it to give credence to controversial decisions. Similarly, Feinstein and Kirchgasler (2015) argue that the public are often uninformed about science and excluded from formulating science-based environmental policies. Wilson and Millington (2013, p. 131) also draw attention to ecological modernization's emphasis on science and engineering in solving sustainability challenges, because it often appears to be the "only" viable response. These ecological modernization approaches downplay the value of qualitative accounts of knowledge, neglect local knowledge and civic science, and jeopardize democratic political practices. The top-down and knowledge-intensive environmental paradigm also overlooks ecological and social injustices.

While the dominant ecological modernization approach emphasizes stakeholder relationships and promotes interactions among diverse actors, the approach offers few insights into power relations among those stakeholders and abusive aspects of corporate partnerships. Further, the ecological modernization approach does not inform the actual capacity of NGO-monitored corporate participation in markets to redress environmental problems. These characteristics of ecological modernization also indicate a weak sustainability model. Boykoff and Mascarenhas (2016) criticize the IOC's Agenda 2020 as consisting of less than fully formed policies, believing that it will not convert sustainability principles into meaningful policies and performances that can bring environmental benefits to host cities of the Olympic Games.

Conclusion

The PyeongChang Winter Games, with its emphasis on a low-carbon footprint and stewardship of nature, were an opportunity for Korea to introduce international environmental standards and reform its environmental governance structure and policies through Olympic events. The Olympic-led environmental advancements in Korea focused mainly on reducing carbon emissions. The transportation, construction, and energy sectors applied green technologies and attempted to reduce emissions. POCOG created education programs for green awareness and stakeholder communication and undertook stewardship of biodiversity, ecology, air and noise pollution, waste, water and sewage. Environmental sustainability in and through the Olympics—under the promises of low-carbon emissions and green growth—failed to stimulate a reflexive and meaningful ecological reform for four main reasons. First, the business-as-usual model of the Olympic Games does not correspond to the environmental contexts of host cities and countries. Second, environmental overconsumption leaves an unjust burden on locals. Third, economic calculation of nature pays selective attention to particular environmental issues, and there are limitations on self-reflection of environmental performance. Lastly, the weak model of sustainability prioritizes the appearance of eco-friendliness, which does not translate into long-term, meaningful environmental benefits.

The environmental hazards of the Winter Games will continue to have an enormous impact on low-income, energy-poor residents and farmers in the rural and mountainous areas of PyeongChang and Gangwon. PyeongChang's eco-modernist environmental policy, planning, and practice demonstrate a lack of justice and equity in terms of recognition, process, procedure, and outcome. Procedural justice is challenged by the sport business-as-usual Olympic festival's top-down approach of environmental governance that is driven by scientific and technocratic elites. Meaningful and beneficial ecological modernization through the Olympic Games cannot be solely shouldered by few elites or host nations/cities. The environmental risks associated with the Olympics are an international issue that needs to be addressed by the global community.

As an alternative to weak versus strong sustainability, some scholars advocate a *just sustainability* model and criticize the dominant, stewardship-focused orientation of environmental sustainability (see Dobson, 1999; Agyeman, 2005). Instead of the current orientation toward environmental sustainability, they advocate transformative or just sustainability that implies a paradigm shift which requires sustainability to take on a redistributive function. "The concept [of] sustainability emerged in large part from 'top-down' international processes and committees, governmental structures, think tanks, and international NGO networks," according to Agyeman (2005, p. 2). He contends that justice and equity must move to center stage in the discourse on sustainability. In order to address the unjust environmental burdens that social minorities often encounter, scholars suggest rethinking "environmental justice" as a concept and an approach, rather than "environmental sustainability." Agyeman's (2013) framework for just sustainability provides a justice- and equity-focused understanding of the term "sustainability" and urges a move beyond a simplified "green" discourse to one that recognizes the role of social and economic inequality in environmental disparities.

Notes

1 O_2 Plus was "the environmental vision suggested by POCOG to avoid, minimize and reduce environmental damage and GHG emissions from the PyeongChang 2018 Winter Olympic Games with the aim to achieve a low-carbon PyeongChang2018" (POCOG, 2015, p. 74). In practice, it refers to purchasing a considerably greater amount of carbon credits than the amount of carbon dioxide produced during the Olympics, resulting in the production of oxygen.
2 Solvay is a transnational corporation that manufactures advanced materials and specialty chemicals, addressing next-generation mobility and improving resource efficiency.

3 The rest of the energy sources (88%) were the usual fossil fuel-based and nuclear energy.
4 ISO 14064-1 details principles and requirements for designing, developing, managing, and reporting organization- or company-level GHG inventories.
5 CBD, informally known as the Biodiversity Convention, is a multilateral treaty, and Korea is a signatory and ratified state member among 196 parties. The main goals of the convention are the conservation of biological diversity, the sustainable use of its components, and the fair and equitable sharing of benefits arising from genetic resources.
6 A real-time integrated information system that manages the entire process of waste discharge to disposal via the internet, which is also known as the Manifest System. It is one of the Korean regulatory systems for the waste trade and resource recycling.
7 A contract in which a company is given full responsibility to plan and build something clients must be able to use as soon as it is finished without needing to do any further work on it themselves.

References

Agyeman, J. (2005). *Sustainable communities and the challenge of environmental justice.* New York: New York University Press.
Agyeman, J. (2013). *Introducing just sustainabilities: Policy, planning, and practice.* New York: Zed Books.
Beder, S. (1999). Greenwashing an Olympic-sized toxic dump. *PR Watch, 6*(2), 2–10.
Boykoff, J. (2014). *Celebration capitalism and the Olympic Games.* New York: Routledge.
Boykoff, J., & Mascarenhas, G. (2016). The Olympics, sustainability, and greenwashing: The Rio 2016 Summer Games. *Capitalism Nature, Socialism, 27*(2), 1–11.
Bridge, G. (2002). Grounding globalization: The prospects and perils of linking economic processes of globalization to environmental outcomes. *Economic Geography, 78*, 361–386.
Bumps, A.G., & Liverman, D.M. (2008). Accumulation by decarbonization and the governance of carbon offsets. *Economic Geography, 84*(2), 127–155.
Buttel, F.H. (2000). Ecological modernization as social theory. *Geoforum, 31*(1), 57–65.
Dobson, A. (ed.) (1999). *Fairness and futurity: Essays on environmental sustainability and social justice.* Oxford, UK: Oxford University Press.
Essex, S., & Chalkley, B. (2004). Mega-sporting events in urban and regional policy: A history of the Winter Olympics. *Planning Perspectives, 19*(2), 201–232.
Ezawa, M. (2015). PyeongChang should take a different path than Nagano. Invited lecture at the Korea National Assembly, Seoul, Korea. February 13.
Feinstein, N.W., & Kirchgasler, K.L. (2015). Sustainability in science education? How the next generation science standards approach sustainability, and why it matters. *Science Education, 99*(1), 121–144.
Findling, J.E., & Pelle, K.D. (eds.) (1996). *Historical dictionary of the modern Olympic Movement.* Westport, CT: Greenwood Press.
Gaffney, C. (2013). Between discourse and reality: The un-sustainability of mega-event planning. *Sustainability, 5*, 3926–3940.
GHG Statistics Korea. (2015). Greenhouse gas inventory. www.gir.go.kr/home/index .do?menuId=37 (accessed April 24, 2018).

Gibbs, D. (2006). Prospects for an environmental economic geography: Linking ecological modernization and regulationist approaches. *Economic Geography, 82,* 193–215.

Giulianotti, R., Armstrong, G., Hales, G., & Hobbs, D. (2015). Sport mega-events and public opposition a sociological study of the London 2012 Olympics. *Journal of Sport & Social Issues, 39*(2), 99–119.

Green Growth Committee. (2009). *Green growth five-year policies (2009–2013).* Seoul, Korea: Presidential Green Growth Committee.

Green Korea United. (2015a). PyeongChang Winter Olympics' environmental problem. *NoxSaekHeeMang,* 243, March/April.

Green Korea United. (2015b). Special topic: Preservation effort of Mt. Gariwang for a year, it's not too late. *NoxSaekHeeMang,* 248, July/August.

Ha, M., & Yoon, G. (2010). Green growth policy as a meta-frame. *Korean Policy Studies Review, 19*(1), 101–126.

Hajer, M.A. (1996). Ecological modernization as cultural politics. In S. Lash, B. Szerszynski, & B. Wynne (eds.), *Risk, environment and modernity: Towards a new ecology,* pp. 246–268. London: SAGE.

Karamichas, J. (2013). *The Olympic Games and the environment.* New York: Palgrave Macmillan.

Kim, J.I. (2018). Energy welfare blind spot measures are urgent. www.gasnews.com/news/articleView.html?idxno=82382 (accessed June 30, 2019).

Kim, K.Y., & Chung, H. (2018). Eco-modernist environmental politics and counter-activism around the 2018 PyeongChang Winter Games. *Sociology of Sport Journal, 35*(1), 17–28.

Lenskyj, H.J. (2008). *Olympic industry resistance: Challenging Olympic power and propaganda.* Albany, NY: SUNY Press.

Lubbers, E. (ed.) (2002). *Battling big business: Countering greenwash, infiltration and other forms of corporate bullying.* Monroe, ME: Common Courage Books.

Mangan, J.A. (2008). Prologue: Guarantees of global goodwill: Post-Olympic legacies—too many limping white elephants? *International Journal of the History of Sport, 25*(14), 1869–1883.

Mol, A.P.J., & Sonnenfeld, D.A. (eds.) (2000). *Ecological modernization around the world: Perspectives and critical debates.* London: Frank Cass.

Mol, A.P.J., & Spaargaren, G. (2002). Ecological modernization and the environmental state. *Research in Social Problems and Public Policy, 10,* 33–52.

OECD. (2007). *Environmental innovation and global markets.* ENV/EPOC/GSP(2007)2/REVI. Paris, France: OECD.

POCOG. (2015). *2018 PyeongChang carbon responsible games: Greenhouse gas inventory for the PyeongChang 2018 Olympic Winter Games.* Government Publications Registration Number 11-B552852-000018-01.

POCOG. (2017). *PyeongChang 2018, furthering benefits to people and nature: PyeongChang 2018 pre-Games sustainability report.* Government Publications Registration Number 11-B552852-000252-01.

Preuss, H. (2013). The contribution of the FIFA World Cup and the Olympic Games to green economy. *Sustainability, 5,* 3581–3600.

Rajkobal, P. (2014). Ecological modernisation and citizen engagement. *International Journal of Sociology and Social Policy, 34*(5/6), 302–316.

Sisaweek. (2014). Daelim Industrial, a "forest destructor" at the PyeongChang Olympic construction site: Illegal logging was caught. www.sisaweek.com/news/article View.html?idxno=32194 (accessed June 30, 2019).

Terret, T. (2008). The Albertville Winter Olympics: Unexpected legacies—failed expectations for regional economic development. *International Journal of the History of Sport*, 25(14), 1903–1921.

Timsheva, O. (2001). Environmental legacy of the Olympic Games. In *Report on the International Olympic Academy's special sessions and seminars*, pp. 116–123. www.ioa.org.gr /wp-content/uploads/2016/09/special-sessions-2001-37938-600-21.pdf (accessed July 20, 2019).

UN CBD. (2014). Press release: Governments meet in republic of Korea to assess progress in implementing global strategic plan for biodiversity. United Nations Decade on Biodiversity. www.cbd.int/doc/press/2014/pr-2014-10-06-cop-12-en.pdf (accessed June 30, 2019).

Wilson, B., & Millington, B. (2013). Sport, ecological modernization, and the environment. In D. Andrews & B. Carrington (eds.), *A companion to sport*, pp. 129–142. Malden, MA: Blackwell.

Yonhap News. (2016, November 24). "Too expensive": Municipalities are reluctant to buy eco-friendly deicers. www.yonhapnews.co.kr/bulletin/2016/11/23/020000000 0AKR20161123133600054.HTML?input=1195m (accessed January 10, 2018).

Yonhap News. (2018, February 21). Olympic ski venue, Mt. Gariwang restoration unexpected ... extensive damage. http://app.yonhapnews.co.kr/YNA/Basic/SNS/r .aspx?c=AKR20180221058300004&did=1195m (accessed July 20, 2019).

York, R., & Rosa, E.A. (2003). Key challenges to ecological modernization theory: Institutional efficacy, case study evidence, units of analysis, and the pace of eco-efficiency. *Organization & Environment*, 16(3), 273–288.

Chapter 8

Of mosquitoes and mega-events

Urban political ecologies of the more-than-human city

Carolyn Prouse

Introduction

In May 2016 more than 100 health experts wrote an open letter to the World Health Organization (WHO) urging that "the 2016 Olympic and Paralympic Games must be postponed, moved, or both, as a precautionary concession ... in the name of public health" (Attaran 2016; see also Associated Press 2016). Their concern weighed less than 2.5 milligrams and had a name with a Latin ring: *Aedes aegypti*. This species of mosquito spreads a particularly virulent strain of a virus known as Zika, which has been linked to abnormal brain development of fetuses – a condition called microcephaly. Not everyone agreed about the dangers of Zika, however, not least the International Olympic Committee (IOC). IOC medical Director Richard Budgett responded to the open letter, saying "there is no justification for cancelling, delaying, postponing or moving the Rio Games" (BBC 2016). A team from Yale School of Public Health argued that the risk to athletes and tourists "is very low indeed" (Boseley 2016) due to the location of most athlete and tourist accommodations, widespread air conditioning of tourist areas, and the time of year. A week later the WHO (WHO 2016b) echoed this sentiment, stating that "cancelling or changing the location of the 2016 Olympics will not significantly alter the international spread of Zika virus," but did urge tourists to steer clear of "cities and towns with no piped water or poor sanitation." The Games proceeded as planned.

In this chapter I focus on the *Aedes aegypti* mosquito and the Zika virus in order to understand how the "non-human world" takes shape through sport mega-events (SMEs). Inspired by Timothy Mitchell's (2002) chapter "Can the Mosquito Speak?", I think of the mosquito as an agent that shapes urbanization dynamics and governance effects, but whose agency is tied up in the power, profit, and political dynamics of mega-events. By methodologically foregrounding the non-human, my purpose is to contribute to a growing field of literature in the sociology of sport that interrogates SMEs' relationships with the so-called natural, non-human, and ecological world – of which mosquitoes and viruses are a part. I draw on recent work in urban political ecology (UPE), specifically, to think through the embodiments of SMEs and the co-constitution of social and natural worlds, or socio-natures, through the hosting of these events. I do so in order to upset social/nature binaries in theories of SMEs, and to point to the extensive and distributed effects that these events have on human/non-human life.

I focus on sport mega-events because they are a crucial process of urbanization in which socio-natures are re-made on a large scale. Not only should they be interrogated for this reason, but SMEs also have the ability to reveal socio-natural urban dynamics that are less visible with smaller events and gatherings. What is particular to sport mega-events that makes this the case? SMEs are simultaneously global, national, and local phenomena:

> They necessarily connect the local to the global because their very exist-
> ence requires a sustained engagement between national and local author-
> ities, supporters and critics, with global networks of capital accumulation
> and circulation, culture and communications, international governmental
> relations, international non-governmental organizations (INGOs), and
> international flows of migration and tourism.
>
> (Gruneau & Horne 2015: 3)

SMEs facilitate the movement of spectators from all over the world while, at the same time, often dispossessing large groups of people located adjacent to stadium construction – people who are moved, or who have to move, else-where. SMEs also, crucially, shape people's affective orientations towards sport; they move people to particular desires and feelings that, when coupled with the modernist notion and often unquestioned good of "sustainability," can make it difficult to critically analyze their socio-natures. As such, a robust theoretical and methodological framework that focuses on SMEs *as* socio-natural reconfigurations can help shed light on their urban dynamics. It can help sports scholars and fans alike understand the multi-faceted, multi-scalar, and inter-species/inter-matter transformations and reverberations SMEs precipitate throughout the urban fabric.

The chapter proceeds as follows. First, I briefly review the different ways that sport sociologists and cultural theorists have engaged the natural world, from sport governing bodies' discourses of ecological modernization to the environmental impacts of sport mega-events to the natures that inhere within sporting bodies. I then turn to urban political ecology as an interdisciplinary field that could make important contributions to the sport mega-events litera-ture through disrupting the "naturalness" of the "environment" and by draw-ing attention to the human/non-human embodiments of these gatherings. In the following section I put these ideas to work by briefly considering two circuits, or circulations, of how *Aedes aegypti* comes to matter to sport mega-events: circulating spectators and circulating dispossessions. I discuss each in conversation with the urban political ecology literature. I close the chapter by outlining three methodological maneuvers inspired by this case which might be helpful for thinking of the political ecological socio-natures of mega-events.

Sport, mega-events, and the "natural" world

Sport scholars have been turning a critical eye to the sustainability agendas of sport governing bodies such as the International Olympic Committee (IOC) and FIFA. The United Nations (UN), through its Sustainable Development Goals, has institutionalized sport as a means to create more environmentally friendly forms of development (see Swatuk, Chapter 2 in this collection; see also Millington, Darnell & Millington 2018). The IOC in particular has gone to great lengths to incorporate discourses of sustainable development into its mandate, evidenced in its Agenda 2020 (see Karamichas, Chapter 6 in this collection). The IOC's sustainability agenda is concerned not only with minimizing harm to the environment, but also with introducing and maximizing more sustainable practices, such as through the planting of trees, the creation of green transportation infrastructure, and the cleaning up of "unproductive" land (Boykoff & Mascarenhas 2016; Gaffney 2013; Millington, Darnell & Millington 2018). The IOC thus frames the Games as active agents in improving sustainability throughout the globe.

Sport sociologists often critique these drives to sustainability as a form of ecological modernization (EM). This ideology, adopted by sport governing bodies like the IOC, "decouples profit and environmental degradation, with economic growth now understood to be an avenue towards environmental protection, rather than plunder" (Millington, Darnell & Millington 2018: 11). EM envisions industrialization as having reached a phase in which technology can be leveraged towards sustainable growth (see Kim, Chapter 7 in this collection; see also Karamichas 2013; Millington, Darnell & Millington 2018; Millington & Wilson 2015). Like most forms of modernization, EM posits an ideological separation of the social world from the natural world, and is premised on the domination or control of the latter by the former. The mega-event is premised on the separation of nature from culture; under the guise of ecological modernization, social actors (such as the IOC) are thought to be able to control and improve an externally cast "environment" through technological initiatives. Indeed, modern sporting practice in general is implicated in this dualism, as athletes strive to dominate their environments by overcoming the "natural" world of land, air, water to move higher, faster, stronger (Gruneau 2017).

Sports scholars and urban geographers also critique SMEs as capitalist enterprises that dominate "nature" through the pursuit of profit. This scholarship points to how sport governing bodies often draw on the marketability of "green" rhetoric to promote their events. Jules Boykoff and Gilmar Mascarenhas (2016) liken this brand of sustainability to other corporate social responsibility endeavors, calling it "greenwashing." With this marketing practice, following geographers Cristina Temenos and Eugene McCann (2012), mega-events amount to a "sustainability fix"; organizers appeal to mainstream environmentalism in order to protect their growth strategies and open new avenues of profit. According to these critics, "sustainability" rhetoric amounts more to shifts in discourse than material change, and is concerned primarily with "manag[ing] ecological dissent or pursu[ing] new accumulation strategies" (While, Jonas & Gibbs 2004: 554), often simultaneously (see also Long 2016). Much of this literature either implicitly or explicitly critiques sport mega-events for their harmful effects on "nature." It views hosting SMEs as enterprises that, by their capitalist pursuit of accumulation and growth, necessarily harm the environment – be it through increased carbon emissions via air travel or the destruction of natural habitat for sporting infrastructure (see, for instance, Gruneau 2017; Hayes & Horne 2011).

These analyses of the fraught modernist and capital-driven sustainability agendas of sport governing bodies are crucial and timely. They have taken a much-needed step towards disrupting nature/culture dualisms by emphasizing how SMEs adopt modernist ideologies and capitalist practices. Yet, in their focus on the domination of nature and destruction of the environment, these analyses often operate within a discursive framework that takes the environment as a given, external to those who plan and enact the events. In other words, some of this scholarship has not taken the next step of disrupting its own nature/culture dualism, particularly when it posits a nature "out there" that mega-events harm, such as a park that is destroyed by a new stadium location. If one of the inherent characteristics of the Games, and indeed modern sport itself, is the separation of nature from culture, human from non-human, I suggest then that our analyses and theory about mega-events should *also* disrupt such modernist dichotomies. If we think of SMEs as capitalist processes that accrue profit for local, national, and transnational elites, then we should also attend to how this capitalist modernity is always predicated on and (re)produces uneven relations amongst people and "nature," always creating new nature/culture hybrids.

There have indeed been interventions in sport sociology and physical cultural studies that work towards disrupting nature/culture dualisms. This has most obviously been the case in research that focuses on the cyborg body of athletes – analyses within critical disability studies and science and technology studies that take the body as mediated via technology, whether it be through prostheses or erythropoietin injections. Physical and cultural theorists are increasingly focusing on the *materiality* of these cyborgian bodies, documenting experiences of concussions

(Ventresca 2018), testosterone testing (Karkazis & Jordan-Young 2018), and amenorrhea (Thorpe 2016), among other concerns.

This work on the materiality of the body tends to be oriented towards the experiences of the sporting participant. Scholars are drawing on "new" materialism literature – such as actor network theory, the Deleuzian assemblage, the Science and Technology Studies (STS) cyborg, and the (neo)Marxian laboring body – to understand how the athlete is produced and how athletics is experienced. For instance, Gavin Weedon (2015) has recently used actor network theory to trouble the category of the "human" in the extreme obstacle race "Tough Mudder." He argues that the forms of camaraderie generated between race participants can only be understood by taking seriously the agency of the non-human world: the mud and the grease on the course and the proteins and water within bodies are as vital to the experience – and the very materialization – of the "athlete" as is any combative ethos. He draws on Bruno Latour to argue that the "social" here is:

> a horizontal or "flattened" ontology, democratic in its scope..This means insisting on the radical distribution of agency as the effect of collaborations, as opposed to being set forth from human intentions: everything is active in cultural-natural-technological collectives, and anything present is therefore potentially agentic.
>
> (Weedon 2015: 445)

Weedon's work is a prime example of a recent attempt to disrupt nature/culture dualisms and to radically re-think ontology in sports studies.

Overall, then, there have been important movements in the sociology and cultural studies of sport and physical activity literature to (re)consider sport and the so-called natural world. Sport sociologists have been keen to understand the non-human effects and exigencies of sport-related activity. Literature on sport mega-events, specifically, has focused on how they harm the environment, even when the organizers of said events promote a sustainability agenda. The foci for these scholars tend to be on the profit-driven contradictions between sustainability rhetoric and capitalist growth, and on critiquing the modernist ideology that guides SMEs' domination of nature. What is at stake in the SME literature is understanding the *effects* of the mega-event on an *external* nature – often called the "environment" – while critiquing the discursive construction of "sustainability." The latter set of scholarship I have briefly discussed – the ever-transforming materialist concern with the athletic participant – is vitally concerned with troubling the body as a coherent category of analysis. This work is beginning to engage a wide variety of theoretical tools and epistemologies to trouble any simplistic nature/culture dualism in how sporting bodies are made, experienced, and understood.

Yet neither set of literatures explicitly speaks to, nor obviously explains, the "problem" of the Zika virus at the 2016 Rio Olympic Games. Zika was not affecting athletic participants (or not to any significant extent), and thus has not been theorized by analysts focusing on the athletic body. Moreover, Zika has not been rendered an "environmental" concern of note to most (if any) SME-sustainability scholars, perhaps because, as a virus, it does not exist as an externalized, at-risk "nature" in the way "the environment" is typically cast. What I propose for understanding Zika at the 2016 Olympic Games – and the threat it posed – is to take the different concerns of these literatures seriously and work towards a new framework for understanding sport mega-events: focusing first on the environmental effects of SMEs (while recasting "environment"), and second, disrupting the category of "human." I do so by honing in on the embodied *non*-sport mega-event participants: spectators, mosquitoes, and viruses. I focus on people, for instance, who have been dispossessed for mega-event constructions and meet the mosquito in new locales, and those who travel across the world with virus-laden blood in order to spectate. Specifically, I am interested in how the mega-event brings into relation new social and material worlds that do not take as a given, or a starting point, the ontological primacy of human actors. How, for instance, does the Zika virus and the *Aedes aegypti* mosquito move and exert a (constrained) agency in relation to the mega-event? What new bodies, governance regimes, and built environments are created at the nexus of these relations? And how are these relations, too, being capitalized on by an ever-expanding pursuit of profit?[1]

I propose that sociology and cultural studies of sport could increasingly engage with the inter-discipline of urban political ecology (UPE) to help answer questions such as these and to disrupt the subject/object dualism inherent to concerns over environmental impact. I do not mean to suggest that no one is doing this in the sociology of sport world.[2] What I would like to do here, though, is propose a more explicit and systematic engagement with many of the theories circulating in critical UPE. Thus, in what follows, I outline three different theoretical contributions of this literature. I then turn empirically to the Zika virus and the mosquito, and discuss what this UPE lens may contribute to understanding the 2016 Olympic Games' relationship with Zika.

Urban political ecology

Urban political ecologists view urbanization processes as mutual imbrications of nature and culture. Urbanization occurs within and through "a vast network of relations, and within complex flows of energy and matter, as capital, commodities, people and ideas, that link urban natures with distant sites and distant ecologies" (Braun 2004: 637). Through this lens, "nature" is not separate from human activity, but produced and constituted through social action, while non-living beings themselves have agency in shaping social worlds. Political ecologists use the term "social nature" or "socio-nature" to refer to these hybridities.

Early urban political ecologists used Marxist theories to think through the co-constitution of the natural and social world. In their view, capitalist processes create the "environment": social labor (generally but not exclusively capitalist) produces the nature of the city; in the words of Neil Smith, nature is "incomprehensible except as mediated by social labour" (Smith 2006: xiii). This labor produces the city as a *metabolic socio-nature*, a "circulation of matter, value and representation" (Smith 2006: xiii) that materializes through concrete, garbage dumps, water infrastructure, and housing. Scholars such as Erik Swyngedouw, Maria Kaika, and Nik Heynen have elaborated on this idea of "city as metabolism" to argue that "All socio-spatial processes are invariably ... predicated upon the circulation and metabolism of physical, chemical, or biological components. Non-human 'actants' play an active role in mobilizing socio-natural circulatory and metabolic processes" (Heynen, Kaika & Swyngedouw 2006: 12). Nature is not only produced through this capitalist mode of production, but also internalized to it: what is often considered "natural" is increasingly commodified, with water, recycling, and food production incorporated into for-profit industries. True to a Marxist perspective, these productions of nature – or socio-natures – take shape through exploitative capitalist relations, in which "societies make the natural environment they live in ... although not of course under conditions of their own choosing" (Smith 2006: xiv).

Recently, urban political ecologists have turned to actor network theory and the Deleuzian assemblage to theorize a decidedly more flattened notion of agency and the city. To political ecologist Bruce Braun, discussing the SARS virus, it is:

> possible to understand cities, for example as "posthuman" assemblages .... [Writing] the virus into a "posthuman" Toronto [for example] explodes the time-space of the city, folding people and animals in China and Thailand into bodies on Queen Street, and revealing time to be multiple and rhythmic – the time of circulation of people and capital but also molecules.
>
> (2004: 273)

This conception of the city is akin to Weedon's theorizations of the obstacle course: the non-human world exerts agency in, through, and apart from "humans," with the (non-)human world networked in a distributed form of power and interrelation.

As these prominent theorizations have taken hold in urban political ecology, so too have modes of thinking that upset the flattened approach of assemblage urbanism, and the capital-centric analyses of urban metabolism.[3] Feminist critiques of urban metabolism have argued that metabolic processes are inherently embodied, and thus scholarship must connect "socio-natures of consumption, waste and resource distribution with the intimate, meaningful and power-laden embodiments of such flows among differently situated groups" (Doshi 2017: 126). Feminist UPE also pays attention to the ongoing histories of social power that shape the urban assemblage or metabolism – specifically asking how heteronormativities, patriarchies, and racializations inform the capitalist metabolic process or the "distributed" agency within an assemblage (Doshi 2017; Heynen 2017). Relatedly, Nik Heynen (2016) has called for an "abolitionist" political ecology that interrogates the production of nature through Cedric Robinson's notion of racial capitalism; he argues that capitalist processes that produce socio-natures are always shaped through ongoing histories of racial valuation, often inscribed in the landscape. To Henyen, an abolitionist political ecology analyzes "how cities have been produced through racialized logics that have been engineered into their building blocks, facades, plumes of dust, streams, forests, and air circulations" (Heynen 2016: 842). Taken together, these latter approaches seek to deepen, enrich, and center various intersecting modes of power (that is, beyond flattened or capital-centric) that shape the production or assemblage of socio-natures.

Overall, this brief discussion of urban political ecology highlights the eclectic and sometimes contradictory modes of analyses present in the interdiscipline. Yet what they all have in common is a troubling of the social/nature dualism at the heart of modernist thought. They do so through upending the very idea of the city: to UPE scholars, the urban is a socio-nature produced or assembled through human and non-human entities that are continuously re-configured through urbanization processes – such as through the building of houses, the distribution of water to homes, and the hosting of sport mega-events. A primary way that UPE scholarship upsets modernist dualisms is to center the non-human agent as it acts through socio-natural hybridities.

What can tracing non-human entities reveal about sport mega-events and their production or assemblage of socio-natures? And how might this methodological approach assist us in theorizing the distributed and often harmful effects of hosting SMEs beyond their domination of an externalized "environment"? In what follows I turn empirically to the Zika virus and the mosquito to understand how new socio-natures are produced through hosting SMEs in Rio de Janeiro. I also focus on the governance regimes that these new socio-natures legitimate, and pay attention to the ongoing histories of poverty, race, and gender relations that shape the production of the hybrid city. I do so through thinking of two routes of circulation: transnational circulations of the virus and spectators, and local circulations of the mosquito and displaced people.

Circulating viruses and vectors

Circuit 1: viruses, mosquitoes, and transnational bodies

The sport mega-event participates in and reshapes urban political ecologies by circulating matter. SMEs facilitate transnational flows of human bodies that are themselves comprised of viruses, among billions of other forms of organic and inorganic material. Scientists have gone to great lengths to understand how the circulation of people has facilitated the travel of Zika, specifically, to Brazil. Microbiologists mapping the virus's genetic code believe that the type of Zika in Brazil is not the virus's original strain, from the Zika forest in Uganda, but is a second, more virulent, strain originating from Asia (FutureLearn 2017).[4] According to these genetic mappings, the virus arrived in the country sometime between late 2013 and mid-2014. Most commentators in Brazil now believe that the virus was brought during either the Va'a World Sprint Championship canoe race or FIFA's Confederations Cup (Arbex et al. 2016). The 2014 FIFA World Cup has also been implicated in bringing Chikungunya infection, hemorrhagic fever, cholera, and avian influenza A (Arbex et al. 2016). According to these studies, mega-events are implicated in (re)constructing human-virus assemblages: they are re-organizations of space-times where biomes and microbiomes, carried in the material substrates of people, entangle and are enmeshed on airplanes, in the streets, and, indeed, in the very body of the mosquito. The virus here is agential: it causes disease and disorder in human bodies, particularly when it re-shapes the skulls and brains of developing fetuses.

Yet the Zika virus does not move through flattened space. Rather, its global mobility is shaped by the ability of some people to travel – requiring resources of leisure time and money – and the advanced technology (and capital) necessary for space-time compression. We can thus think of transnational virus transmission through a Marxist production-of-nature lens. The multi-billion-dollar air travel industry has produced new socio-natures of the Zika virus in two major ways: first, it has shifted the spatiality of bodies (and the biological/viral materials they house), bringing into new proximate relation people from diverse locales. Second, and more indirectly, airplane technologies have large carbon footprints that are implicated in climate change – itself a kind of socio-nature – which shifts the pattern of mosquito mobilities. The warming of the earth, for instance, has seen *Aedes aegypti* mosquitoes migrate north, and put new countries and continents at risk of the viruses they may carry

(FutureLearn 2017). SMEs are thus directly and indirectly implicated in the shifting spatialities of humans, viruses, and mosquitoes as they produce ever-transforming socio-natures.

These new encounters between different forms of matter – such as disease-causing vectors and humans – inspire new governance regimes intended to manage these "risky" socio-natures, particularly when they put more privileged bodies at risk. For instance, six months prior to the Rio Olympics, in February 2016, the World Health Organization declared Zika a Public Health Emergency of International Concern (PHEIC) (Ventura 2016). Scholars in Brazil have argued that this public health crisis was called only because the Olympic Games were putting 500,000 tourists and athletes from *elsewhere* at risk. PHEICs are declared in order to "prevent the disease from leaving the place where it should have stayed" (Ventura 2016: 3) – namely, most commonly, in Southern countries. As one journalist put it "In this case, Zika has become a global emergency by threatening the brains of children from rich countries" (Brum 2016).

Zika virus transmission is also a reproductive encounter between a "maternal" body and fetus, and therefore inspires a gendered and sexualized framework of governance. The WHO indeed issued an international travel advisory, cautioning "pregnant women not to travel to areas with ongoing Zika virus transmission" (WHO 2016a), and advised women in Latin America not to get pregnant. Taking this advice a step further, high-profile athletes such as golfer Rory McIlroy decided to forgo the Olympic Games to protect not himself, but the future reproductive health of his partner, to whom he did not want to sexually transmit the virus. Yet it is poor, racialized women, often single mothers, who have been most affected by Zika, and who are made responsible, through these advisories, for potential future generations of babies with microcephaly – responsibility that comes without access to adequate health care to manage virus/fetus encounters (Cavalcante 2016). Sport mega-events thus bring into new assemblage peoples and viruses from all over the world, but these encounters are shaped by differential geographies of precarity and value – who has the means to travel; who, based on reproductive assumptions, is advised not to travel; and who is made responsible for future microcephaly without adequate support. Crucially, then, the Rio Olympics assembled different matter – they produced new socio-natures of viruses in human bodies – but also instituted new (gendered) governance regimes based on whose *bodies* matter.

On the one hand, then, we can think of how mega-events facilitate new encounters between human and non-human entities through transnational movement, and institute new public health governance regimes shaped through power-laden geographies of mobility and access. But mega-events also actively produce new socio-natures of the urban as they displace both people and policy. I briefly pursue these themes – the circulations of local people, policy, and mosquitoes – in the following section.

Circuit 2: mosquitoes and displaced peoples

Like other mosquitoes that carry disease, the *Aedes aegypti* lays its eggs in standing water. Unlike other mosquitoes, however, the Zika-carrying vector prefers clean water. Yet standing clean water, alongside dirty water, is more likely to be found in precarious settlements, often called favelas, in Brazil. Many favela communities have water infrastructure, yet generally the resource is not predictably delivered. Thus, when water is flowing into the community, residents often pump it into large water storage containers on their roofs. If not properly covered, the containers make ideal mosquito breeding grounds (Watts 2016). More-over, many of these communities also lack adequate sanitation and drainage infrastructure and thus periodically flood, leaving pools of standing water between densely packed homes. In poorer areas people are also less likely to afford protections against the mosquito, such as air conditioning and insect repellent. As such, many Brazilian commen-tators are calling Zika a "disease of the poor" (Arbex et al. 2016), even if the *Aedes aegypti* mosquito prefers clean water. This spatiality of the disease is reflected in the aforementioned WHO travel advisory for the Olympic Games, which cautioned tourists and athletes to stay out of areas with "poor sanitation." Risk of Zika transmission, then, reveals a historically entrenched geography of differential precarity, shaped through ever-transforming socio-natures of infrastructure, water, mos-quitoes, and poverty.

Sport mega-events remake this "nature" of the city in both indirect and direct ways. SMEs, in general, reshape the metabolism of the city: they shift flows of capital, flows of water, flows of heat, and flows of mosquitoes, often through their associated infrastructural development. Indirectly, SMEs participate in urban metabolic flows via processes of gentrification: they help valorize regions and communities generally located adjacent to new stadiums, facilitating the building of infrastruc-ture and thus protecting areas from disease-bearing vectors. As capital flows through the environment, other regions of the city are de-valued, where infrastructure becomes neglected, falls into disrepair, and increases human susceptibility to infectious disease and toxins (Ranga-nathan 2016). These new socio-natures are inherently embodied, as infrastructural decisions shape the flow of viruses through bodies and lives of differential value.

Sport mega-events are also directly involved in producing socio-natures through displacement of people and policy. For instance, Cariocas were actively displaced from their homes in communities adjacent to stadium construction. They were relocated to, or found new homes/accommodations in, areas of poor infrastructure that have allowed *Aedes aegypti* to flourish. According to Sá, Reis-Santos and Rodrigues (2016: e603):

> Over 50,000 people were forced to move from informal settlements in areas with reasonable infrastructure to the outskirts of the cities.. They will join many already living in poor areas – where sanitary conditions and waste management are worse – increasing the proportion of the population forced to store water and the amount of garbage thrown in water streams, blocking water flow, both of which favour the proliferation of *A aegypti*, the vector of Zika, dengue, and chikungunya viruses.

Here, the displacement of people, for the capital accumulation of the IOC, private construction firms, and other corporate entities, manifests in new socio-natures of (constrained) mobility: poor and racialized populations are forced to move to already-underserved communities, thus exerting pressure on depleted infrastructure, worsening sanitation conditions, and increasing the risk of virus transmission. In other words, the mega-event creates new metabolic flows of people and mosquitoes that increases embodied precarity.

Sport mega-events also affect infrastructural development – and thus the socio-nature of mosquitoes, viruses, and human bodies – through the displacement of policy. Rio de Janeiro had a progressive community-led upgrading project, called Morar Carioca, that would have addressed many sanitation issues in the city through a sustained multi-sectoral approach. Brazilian urbanist Mariana Cavalcanti called Morar Carioca an urban planners' dream (Steiker-Ginzberg 2014), centered on full community participation and integrated service delivery. Infrastructural upgrading – targeting sanitation, among other areas – was a significant arm of the program, and would have stifled the movements of *Aedes aegypti* through favelas. Yet the project was largely disbanded by Rio's then-mayor Eduardo Paes for the immediate capital and spatial requirements of the FIFA World Cup and the Olympic Games. In other words, this policy to upgrade the infrastructure of low-income settlements was displaced for capital investment in stadiums and transportation. As such, the SMEs compounded already-existing infrastructural neglect that has allowed mosquitoes – and thus infectious disease – to flourish.

Overall, my purpose here has been to center the mosquito and the Zika virus in ever-transforming infrastructural environments. Using this approach, one can begin to see the extensive and diverse ways that sport mega-events produce new socio-natures. In this section I have focused on the movement of capital, policy, and people to give a sense of an SME's geographic reverberations and space-time metabolic flows. As in the last section, my brief discussion here demonstrates that these flows do not occur through a flattened ontology; rather, they take shape through histories of differential precarity informed by race, gender, and class relations. While I have not been able to do justice to the expansive metabolisms or socio-natural networks of SMEs, these empirical sections have begun to demonstrate what a focus on viruses, mosquitoes, people, and capital might offer for understanding the socio-natural transformations associated with hosting.

Towards an urban political ecology of sport mega-events

Above, I have put into conversation a number of diverse literatures, from sport sociology of mega-events to Brazilian public health analyses to theories of urban political ecology. I have explored some of the transformations effected by the 2016 Olympic Games, alongside other SMEs hosted in Brazil, by foregrounding the new encounters they have produced between the mosquito, the Zika virus, and the non-sporting participant. Drawing on this case and UPE scholarship, I next suggest three epistemological maneuvers through which we might (re)think the political ecologies of mega-events.

Human/non-human boundaries

In this chapter I have briefly explored how SMEs facilitate the flow of the Zika virus, the mosquito, and various bodies through transnational and regional space. By focusing on similar encounters between the human and non-human world, critical theorizations of SMEs could further contest the modernist boundaries of the human/non-human world, exploring how these boundaries are actively and continuously permeated, undone, and redone in the making of what we might call the urban. This means troubling binaries between human and non-human (bodies, viruses, infrastructures), as well as boundaries between differential valuations of humans (that is, who is at risk of public health edicts and capitalist developments; who is invisibilized and erased; and who is surveilled and governed and via what technologies). I have presented a few different insights from the eclectic inter-discipline of urban political ecology that may help us think differently about the socio-natures of SMEs: assemblage urbanism approaches that focus on the agency of the non-human, the capitalist production of urban environments as socio-natures, and the embodied racial, classed, and patriarchal histories that continue to shape socio-natural hybridities. These approaches have different philosophical foundations, yet all point to how the "social" and "natural" world are never separate and are shaped through ongoing relationalities.

Space-time analyses

Sport mega-events are influenced by, and create new flows beyond, the site and moment of the event. I suggest that we follow humans, mosquitoes, and viruses as they are being reconfigured in SME environments through uneven ecological development. Because capitalist modernities depend on invisibilizing particular peoples and "natures," it is necessary to follow the flows of people and things – such as travelers and displaced people – historically and contemporarily as well as geographically. The Zika virus, in particular, also points to the necessity of focusing on *reproductive* pasts and futures. With Zika, the effects of uneven ecological development may be displaced temporally (Murphy 2017) onto the bodies of babies yet to be born. This displaced temporality has implications for public health governance strategies that regulate reproductive bodies and their potential futures – of those who are or might become pregnant.

Interscalar dynamics

Finally, this case has hinted at the interscalar nature of SMEs. These events are articulations of policies and events happening at many different scales: for instance, the IOC's sustainability goals repeatedly fail because of their constrained timelines and capitalist accumulation strategies, but also because of municipal and national programs and policies that have their own historical dynamics and regimes of power. A crucial point of analysis in any political ecology approach is the interrelatedness of these scales. But these analyses should also center the scale of the (non-sporting) body: how differences of precarity are embodied along intersecting lines of power, and how these differences are lived and experienced in daily life as mega-events foment new ecologies within and across diverse peoples.

Conclusion

Engaging political ecological frameworks throws into question the very idea of sustainability as protecting an externalized nature. Indeed, as long as international sports governance and hosting bodies posit sustainable development through the lens of the environment, there will always be invisibilized "externalities" that do not fit hegemonic notions of said environment – such as Zika virus transmission. The implication is that sustainable initiatives will continue to fail the peoples and "natures" who are always-already made precarious and invisible through (neo)colonial and capitalist development. In other words, the common visions of sustainable development will continue the modernist maneuver of separating nature from culture that is at the heart of colonial-capitalist destruction.

Overall, the urban political ecology approach I propose here can help us understand the temporal and spatial effects of mega-events. Crucially, it might allow scholars to theorize SMEs' re-workings of socio-natures beyond a focus on an externalized "nature" or "environment." These analyses are indeed already beginning in the field. What I have hoped to offer here is an impetus for sport sociology and physical cultural studies to engage more systematically with the exciting sub-discipline of urban political ecology. Doing so requires decentering the human and taking seriously active agents that affect life chances and vulnerabilities. This theoretical and methodological maneuver offers a fuller perspective of how SMEs remake our urban worlds.

Notes

1 Admittedly, my questions do not completely de-center a concern with "human sovereignty" (Weedon 2015); I am asking how mega-events and the Zika virus have a disproportionate effect on specific populations of humans. Yet I do so in ways that, I think, might offer a more nuanced understanding of the socio-natures of sport mega-events.
2 For instance, Millington, Darnell & Millington 2018 have engaged with political ecologist Paul Robbins's work, and Millington and Wilson (Chapter 3 in this collection) are increasingly considering the political ecology of golf courses.
3 Of course, decolonial, critical race, and feminist epistemologies have long questioned the nature/culture dualism. My point here is that this thinking is becoming increasingly codified and recognized – if not always engaged – in more hegemonic urban political ecology.
4 This account of Zika relies on an assumed "truthfulness" to the always-politicized science that produces it, which a thorough interrogation of Zika would need to discuss. Given the limited scope of this chapter, however, I will not delve into an STS-inspired analysis here.

References

Arbex, Alberto, Bizarro, Vagner, Paletti, Mikele, Brandt, Odirlei, de Jesus, Ana, Werner, Ian, Dantas, Luiggi & de Almeida, Mirella (2016). Zika virus controversies: Epidemics as a legacy of mega events? *Health* 8: 711–722.

Associated Press. (2016). Health experts: Move Rio Olympics due to Zika outbreak. *CBS News*. Retrieved 20 April 2017 from www.cbsnews.com/news/health-experts-urge-who-to-move-postpone-rio-olympics-zika-virus

Attaran, Amir (2016). Off the podium: Why public health concerns for global spread of Zika virus means that Rio de Janeiro's 2016 Olympic Games must not proceed. *Harvard Public Health Review 10*. Retrieved 20 April 2017 from http://harvardpublichealthreview.org/off-the-podium-why-rios-2016-olympic-games-must-not-proceed/

BBC. (2016). Olympics 2016: IOC insists games will go ahead despite Zika. *BBC World News*. Retrieved 24 April 2017 from www.bbc.com/news/world-latin-america-36272502

Boseley, Sarah (25 July 2016). Zika virus risk at Rio Olympics "negligible", says Yale report. *The Guardian Online*. Retrieved 21 April 2017 from www.theguardian.com/world/2016/jul/25/zika-virus-risk-at-rio-olympics-negligible-says-yale-report

Boykoff, Jules & Mascarenhas, Gilmar (2016). The Olympics, sustainability, and greenwashing: The Rio 2016 Summer Games. *Capitalism Nature Socialism* 27(2): 1–11.

Braun, Bruce (2004). Querying posthumanisms. *Geoforum* 35: 269–273.

Brum, Eliane (2016). The Zika virus mosquito is unmasking Brazil's inequality and indifference. *The Guardian Online*. Retrieved 25 April 2017 from www.theguardian.com/commentisfree/2016/feb/16/zika-mosquito-brazil-inequality-brazilian-government

Cavalcante, Thais (2016). Rio Olympics: View from the favelas – "I've seen six people infected with Zika. I am one of them". *The Guardian Online*. Retrieved 30 April 2017 from www.theguardian.com/global-development/2016/mar/30/rio-olympics-view-favelas-six-people-with-zika-virus

Doshi, Sapana (2017). Embodied urban political ecology: Five propositions. *Area* 49 (1): 125–128.

FutureLearn. (2017). Preventing the Zika Virus: Understanding and controlling the Aedes mosquito. *FutureLearn: Online Course*. Retrieved 8 May 2017 from www.futurelearn.com/courses/preventing-zika#section-overview

Gaffney, Chris (2013). Between discourse and reality: The un-sustainability of mega-event planning. *Sustainability* 5: 3926–3940.

Gruneau, Richard (2017). *Sport and Modernity*. Boston, MA: Polity Press.

Gruneau, Richard & Horne, John (2015). Mega events and globalization: A critical introduction. In Richard Gruneau & John Horne (Eds.) *Mega-events and Globalization: Capital and Spectacle in a Changing World Order* (pp. 1–44). London: Routledge.

Hayes, Graeme & Horne, John (2011). Sustainable development, shock and awe? London 2012 and civil society. *Sociology* 45(5): 749–764.

Heynen, Nik (2016). Urban political ecology II: The abolitionist century. *Progress in Human Geography* 40(6): 839–845.

Heynen, Nik (2017). Urban political ecology III: The feminist and queer century. *Progress in Human Geography* 42(3): 446–452.

Heynen, Nik, Kaika, Maria & Swyngedouw, Erik (2006). *In the Nature of Cities: Urban Political Ecology and the Politics of Urban Metabolism*. New York: Taylor & Francis.

Karamichas, John (2013). *The Olympic Games and the Environment*. New York: Palgrave Macmillan.

Karkazis, Katrina & Jordan-Young, Rebecca (2018). The treatment of Caster Semenya shows athletics' bias against women of colour. *The Guardian Online*. Retrieved 26 April 2018 from www.theguardian.com/commentisfree/2018/apr/26/testosterone-ruling-women-athletes-caster-semanya-global-south

Long, Joshua (2016). Constructing the narrative of the sustainability fix: Sustainability, social justice and representation in Austin, TX. *Urban Studies* 53(1): 149–172.

Millington, Brad & Wilson, Brian (2015). Golf and the environmental politics of modernization. *Geoforum* 66: 37–40.

Millington, Rob, Darnell, Simon & Millington, Brad (2018). Ecological modernization and the Olympics: The case of golf and Rio's "Green" Games. *Sociology of Sport Journal* 35: 8–16.

Mitchell, Timothy (2002). *Rule of Experts: Egypt, Techno-politics, Modernity*. Berkeley, CA: University of California Press.

Murphy, Michelle (2017). *The Economization of Life*. Durham, NC: Duke University Press.

Ranganathan, Malini (2016). Thinking with Flint: Racial liberalism and the roots of an American water tragedy. *Capitalism Nature Socialism 27*(3): 17–33.

Sá, Thiago, Reis-Santos, Barbara & Rodrigues, Laura (2016). Zika outbreak, mega-events, and urban reform. *The Lancet: Correspondence 4*: e603.

Smith, Neil (2006). Foreword. In Nik Henyen, Maria Kaika & Erik Swyngedouw (Eds.) *In the Nature of Cities: Urban Political Ecology and the Politics of Urban Metabolism* (pp. xi–xv). London: Taylor & Francis.

Steiker-Ginzberg, Kate (2014). Morar Carioca: The dismantling of a dream favela upgrading program. *Rio on Watch: Community Reporting on Rio*. Retrieved 12 January 2015 from www.rioonwatch.org/?p=17687

Temenos, Cristina & McCann, Eugene (2012). The local politics of policy mobility: Learning, persuasion, and the production of a municipal sustainability fix. *Environment and Planning A 44*: 1389–1406.

Thorpe, Holly (2016). Athletic women's experiences of amenorrhea: Biomedical technologies, somatic ethics, and embodied subjectivities. *Sociology of Sport Journal 33*: 1–13.

Ventresca, Matt (2018). The curious case of CTE: Mediating materialities of traumatic brain injury. *Communication and Sport 7*(2): 135–156.

Ventura, Deisy (2016). From Ebola to Zika: International emergencies and the securitization of global health. *Cadernos De Saúde Pública 32*(4): 1–4.

Watts, Jonathan (7 February 2016). Brazil's sprawling favelas bear the brunt of the Zika epidemic. *The Guardian Online*. Retrieved 23 April 2017 from www.theguardian.com/world/2016/feb/07/brazil-rich-zika-virus-poor

Weedon, Gavin (2015). Camaraderie reincorporated: Tough Mudder and the extended distribution of the social. *Journal of Sport and Social Issues 39*(6): 431–454.

While, Aidan, Jonas, Andrew & Gibbs, David (2004). The environment and the entrepreneurial city. *International Journal of Urban and Regional Research 28*(3): 549–569.

WHO. (2016a). WHO public health advice regarding the Olympics and Zika virus. Geneva, Switzerland: World Health Organization. Retrieved 5 May 2018 from www.who.int/en/news-room/detail/28-05-2016-who-public-health-advice-regarding-the-olympics-and-zika-virus

WHO. (2016b). WHO public health advice regarding Zika virus. Geneva, Switzerland: World Health Organization. Retrieved 23 April 2017 from www.who.int/mediacentre/news/releases/2016/zika-health-advice-olympics/en/

Chapter 9

Neoliberalism and competitive nationalism

What the 2010 Commonwealth Games and 1982 Asian Games reveal about India's development trajectory

Mitu Sengupta

Introduction

Competition to host sport mega-events (SMEs), such as the Olympic Games and FIFA World Cup, remains vigorous, especially among developing countries. As such, the study of the impact of SMEs on host cities and local populations has generated a substantial body of literature. An important finding in this literature is that the socio-economic and environment impacts produced by SMEs – both intended and unintended – are not accidental by-products, but are rather the inevitable outcomes of SME hosting. Black and Northam (2016) argue, for example, that SMEs create "a moment of opportunity for social re-imaginings and redirection that, if productively harnessed, can make exceptional projects possible" (p. 437), and that "there is a tendency for certain countries and cities to make a habit of mega-event hosting a central feature of their development policies" (p. 440).

If SMEs are thus policy instruments, it is possible, at least in theory, to use them to promote ethical development objectives, such as to build the type of inclusive, resilient, sustainable cities that are recommended by the United Nations Sustainable Development Goals (SDGs). As examples, the SDGs outline the contribution that sport events can make to contributing to local economies (SDGs 1 and 8), tackling food waste (SDG 2), reducing pollution (SDG 6), promoting the use of renewable energy (SDG 7), fostering sustainable industrialization and sustainable cities (SDGs 9 and 11), and ultimately helping to combat climate change (SDG 13) while protecting ecosystems (SDGs 14 and 15) (United Nations 2016). Unfortunately, the shape that SMEs have taken on the ground in developing countries suggests a disconnect between policy and practice. While it is conceded that some SMEs have produced socially beneficial consequences – such as when the athletes' village constructed for the 1972 Munich Olympics was converted into community housing for students – the literature points to a stronger emphasis on the adverse consequences of SME hosting in developing countries, including slum demolitions that disrupt poor people's lives and livelihoods, widened socio-economic inequalities and spatial segregation between rich and poor, and weakened local democracy (see, for example, Baviskar 2014; Bénit-Gbaffou 2008; Cornelissen 2011; Dupont 2013; Matheson and Baade 2004; Zirin 2014). The harmful environmental impacts of SMEs is also of concern. Indeed, while environmental sustainability has become an important aspect of global discourses on what we want from cities and from 'development' more generally – as reflected not only in the text of the SDGs, but also in sporting world initiatives such as the South African World Cup's Green Goal Programme 2010 and the International Olympic Committee's Sustainability Strategy – there is concern that the model of change actually followed by SMEs remains stuck in an old groove, in which fast-track economic growth is ranked above all other objectives (see Boykoff and Mascarenhas 2016; Death 2011).

Blame for the dissonance between the lofty ideals of SMEs and the realities on the ground is typically placed at the door of 'neoliberalism,' namely, the ideology and policy model in which market competition is valued above all else in order to spur greater economic efficiency, better income distribution, enhanced individual freedom, and democracy (see Harvey 2006). The critical literature on SMEs points to neoliberal ideals as the motivating force behind bids to host such events, and the cause of much of the resultant deleterious impacts on host cities (see, for example, Pillay and Bass 2008; Silk 2014). In this chapter, however, I identify an ideological culprit beyond neoliberalism that is often overlooked: competitive nationalism. As Law and Mooney (2012, p. 66) argue, competitive nationalism mobilizes the resources of the state to 'attract mobile global investment and secure the accumulation of [I]ndigenous capital [... to] offer various capital-friendly incentives, including lower relative wage costs, flexible labour supply, and infrastructural requirements through higher public spending.' 'In the long run,' Law and Mooney (2012, p. 66) continue, 'competitive nationalism projects that private capital will raise the absolute standard of living even if wealth and income distribution become grossly unequal within and across regions.' I suggest that competitive nationalism ties SMEs to a model of development, focused on large-scale 'modernization' projects, that is fundamentally resistant to implementing calls for environmental sustainability, improved social impact, greater citizen participation, human rights, and other progressive discourses propagated within the global civil sphere by organizations ranging from the United Nations and the International Olympic Committee, to the International Monetary Fund and the World Bank.

In particular, I engage in a comparative study of two SMEs that were held in Delhi, India's national capital, at markedly different points in the country's development trajectory – the 1982 Asian Games, which occurred while India was still following a state-led model of development, and the 2010 Commonwealth Games, which were held almost 20 years after it had definitively liberalized its economy. I suggest that the reason why SMEs ultimately fail to align with shifts in global understandings of 'development' lies not in the power of neoliberalism *per se*, but in the resilience, in developing countries such as India, of an elite-driven model of development that is focused almost exclusively on boosting economic growth and, more broadly, on 'catching up' with the economic and technological achievements of advanced industrial states. As I argue below, the deleterious consequences of this model of 'catch-up' development, which has its moorings in modernization theory, have certainly been *intensified* by neoliberalism (see Darnell and Millington 2015). However, the ideological origins of the model of development followed in India lie in competitive nationalism, which predates the advent of neoliberalism and continues to be a major force in emerging powers. Because SMEs tend to be of enormous symbolic significance to their host countries, they are particularly susceptible to being utilized by elites to promote 'catch-up' development.

Modernizing India: from statism to neoliberalism

When India obtained independence from British colonial rule in 1947, its government, led by Prime Minister Jawaharlal Nehru, adopted a model of economic development that aimed for modernization and high economic growth through state-directed industrialization (see Kohli 2004). India's choice of development strategy and the logic of its national planning process were greatly influenced by ideas that were dominant within the discipline of 'development economics' at the time.[1] Conversely, development theory was also "for a while greatly influenced by the Indian case" (Chakravarty 1987, p. 4). A prevailing argument during this period was that although the precepts of neoclassical economics were applicable to the analysis of problems in advanced industrial states, they were not relevant to the study of developing countries. It was said, among other things, that 'market failures' had locked developing countries into a vicious cycle that could be broken only by a strong interventionist state and generous international assistance (Myrdal 1957; Nurske 1953). Unlike 'late developers' such as Japan and Germany, which could draw upon the capital, technology, and experience of early developers such as Britain and France, the new crop of low-income countries were too far behind or 'backward' to catch up with the rest unless a dominant state provided a 'big push' that would help break through domestic and international barriers to successful development (see Gerschenkeron 1962; Rosenstein-Rodan 1943). An interventionist state was also viewed as key to achieving equitable and balanced development, since it was thought that wholesale reliance on the market mechanism would lead to excessive consumption by upper income groups and underinvestment in basic goods and physical infrastructure (Chakravarty 1987, pp. 9–10).

For Nehru, who was influenced by the ideals of socialism, the potentially redistributive function of a strong state held special appeal. His espousal of a state-directed development model for India did not, however, go unopposed. Conservative alliances, both inside and outside his party, wanted freer markets and more room for private enterprise. Nonetheless, the primary objective of Nehru's strategy – of creating a strong, modern, and self-reliant India – was supported across the board, by both the left and right (on this, see Khilnani 1997). The symbols of such strength and self-reliance – the 'temples of modern India,' as Nehru described them in a speech inaugurating the Bhakra Nangal Dam (Wyatt 2005) – were large-scale development projects such as steel mills, power plants, and most crucially, giant hydroelectric dams, whose harmful social and environmental impacts have now become well known (see, for example, Mathur 2013).

Nehru's model of state-directed development was followed, in more or less unchanged form, for more than four decades. Despite several liberalizing episodes along the way, a decisive break with the past only came in July 1991, when the Indian government introduced a cluster of changes in economic policy that were based upon the principles of neoclassical economics (see Joshi and Little 1996).[2] Introduced in response to a severe economic crisis, these liberalizing changes followed, in design and intent, standard prescriptions for 'structural adjustment' that were being advocated by international financial institutions such as the International Monetary Fund and World Bank through the 1980s and 1990s. While many of the intended reforms were only half-heartedly implemented – for example, subsidies to agricultural producers continued to proliferate – government after government, since 1991, persevered with the general thrust of the policy framework adopted by Rao (see Nayar 2001). The persistence of the model and concomitant faith in free markets has led critical scholars such as C.P. Chandrashekhar and Jayati Ghosh (2002) to identify the 1991 reforms as India's turn to 'neoliberalism,' in which the driving objective was to reverse the historic role of the state as India's engine of development. Now in its third decade, India's neoliberal turn has significantly altered the state's relationship with the domestic private sector and international markets, and has also left a deep imprint on India's cities, setting in motion a whole new set of negative social and environmental consequences. It is to this topic that I turn next.

Segregating cities and 'neoliberal urbanism'

Critical reflection on the path of urbanization followed in Indian cities post-liberalization has emerged from a variety of disciplinary perspectives.

Sociologists, for example, have explored the impact on cities of the rise of a politically assertive and urban-based 'new middle class' that is invested in the process of economic liberalization initiated in 1991. Unlike the genteel middle classes of the past that were drawn mainly from the government bureaucracy, India's new middle classes are depicted as connected mainly to the private sector and, through family ties, to affluent Indians living in developed countries (see Baviskar and Ray 2011; Brosius 2010). Brosius (2010), Bhan (2009), and Menon-Sen (2010), among others, have documented the growth of anti-poor activism by this new middle class, and how it strategically organizes itself into citizens' groups, such as Residents' Welfare Associations, to mount campaigns to 'cleanse' upscale neighbourhoods of hawkers and vendors, and to enclose and securitize public land for their own private consumption. It is noted, furthermore, that this new middle class has a weaker sense of social responsibility towards the poor than the well-to-do of Nehru's time, who, despite their paternalistic outlook, still viewed poor people as deserving of state-sponsored measures of social welfare. Menon-Sen suggests, for example, that there is 'very little sympathy [for poor people] among more privileged citizens, whose commitment to the old values of charity, social solidarity, and concern for the less fortunate can no longer be taken for granted' (2010, p. 681).

Open contempt for the poor is evident in the way that old cities are being recreated and new cities are being planned. For example, gated communities and private townships have proliferated. They are separated from their surrounding environment by fences, razor wire, and 24-hour video surveillance, and the only connection their occupants have with poor people is through service staff – the armies of maids, nannies, drivers, and cooks who filter in and out of heavily guarded gateways. Unlike the more socially mixed neighbourhoods of the past, however, when domestic workers lived within or close to the homes of their employers, the service populations of post-liberalization India are mostly commuters. They reside in the peripheries of the city, in densely populated informal settlements (or 'slums'), and are obligated to travel long hours to their places of work (see Chatterjee 2004, pp. 131–148).

Another salient view, in the critical literature on post-liberalization Indian cities, is that state institutions, with the judiciary in the lead, consistently discriminate against the urban poor, even in the face of considerable opposition from slumdwellers' organizations and pro-poor civil society groups. One example is Goldman's study of the rise of Bangalore as a 'global city,' which, he suggests, has been facilitated by subsidized land and tax holidays awarded to high-end information technology firms by the government. Goldman argues that "land speculation and active dispossession inside and surrounding the city of Bangalore is the main business of its government today" (2011, p. 557), and also that government actors are "deeply entangled with real estate brokers/dealers and political party leaders" in auctioning off public lands and goods (p. 576). Social relations that are already tense have been further exacerbated, he says, as "public space and lands shrink in size, use and access" (p. 574). What is worse, Goldman suggests, is that much of the land acquired in the name of public interest has not been used for the development of infrastructure – promised expressways and runways for the airport – but for "high value gated residential communities, seven-star hotel complexes, 'medical tourism' hospitals, and business centres" (p. 569).

In relation to the judiciary's anti-poor outlook, Bhan's study of slum evictions in Delhi is instructive. Bhan indicates that between 1990 and 2003, 51,461 low-income houses were demolished in Delhi under 'slum clearance' schemes (2009, p. 128). The pace picked up between 2004 and 2007, when another 45,000 homes were demolished. Bhan indicates that these evictions were the result of rulings in favour of public interest litigations filed by 'non-poor resident welfare and trade associations' in Delhi courts (2009, p. 128). He suggests that while informal settlements and evictions are not new to Delhi, the evictions of post-liberalization Delhi "are different, not just in degree but in kind, from evictions in the past." The key difference, he says, lies not just in increased frequency and intensity, but in the involvement of courts and "the use of altered definitions of 'public interest' [...] [wherein] evictions [are] successfully framed as just, ethical and markers of good governance" (p. 128).

In the 1980s, Bhan points out, court judgments in favour of evictions would insist upon finding alternative accommodation for potential evictees in advance of their eviction, thereby showing at least some concern for the welfare of pavement dwellers and slum dwellers. In recent years, however, court rulings have shown little interest in resettlement and compensation, and the "conception of self-government through market participation [...] has replaced the notion of state patrimony" (p. 137). In other words, poor people are expected to fend for themselves by participating in the market, which places 'a new burden of responsibility upon them, but with little or no change in their capacity to bear it' (p. 138). Bhan's understanding is shared by Goldman, who suggests that the 'active dispossession' at work in Bangalore has been justified on ethical grounds:

> This urban modernization approach argues that rural peasants need not continue to play the static role of urban salesmen, pushing old handcarts selling pitiful bundles of vegetable and herbs [...]. The poor should expect more from their cities, and these globally integrating projects are precisely the vehicle to meet such expectations.
>
> (2011, p. 560)[3]

An important insight to be gleaned from critical urban theorists is that the shifts observable in India are not unique, but are part of broader transformations occurring at the global level. The pattern of urban change perceptible in India is illuminated and provided with global context, for instance, by David Harvey's concept of 'accumulation by dispossession' (2004), and that of 'neoliberal urbanism,' as articulated, for example, by Peck, Theodore, and Brenner (2009). The concept of 'accumulation by dis-possession' – developed by Harvey to understand the process through which public assets are emptied into the market, prying open new avenues for the advancement of capitalist development – has been used extensively in the study of Indian cities. It is used by Goldman (2011), for example, to theorize the seizure and conversion of public land in Bangalore into securitized private enclosures. Peck, Theodore, and Brenner's list of 'mechanisms of neoliberal urbanism' – such as 'devolution of tasks and responsibilities to municipalities,' 'privatization and outsourcing of muni-cipal services,' 'creation of new opportunities for speculative investment in central-city real estate markets,' and 'diffusion of generic [...] approaches to "modernizing" reform among policymakers in search of "quick fixes" for local social problems (e.g., workfare programs, zero-tolerance crime policies)'(2009, pp. 59–62) – is also applicable to many Indian cities, as indicated by Goldman (2011), Bhan (2009), and others.

The 1982 Asian Games: looking behind neoliberalism

Although smaller in scale than contemporary SMEs, the 1982 Asian Games – hosted by India while it was still following a state-directed model of economic development – left a large imprint on Delhi and did so in similar shape and substance to the eventual 2010 Commonwealth Games (CWG). Slum demolitions, for example, cleared the way for the building of sports facilities, and aggressive 'beautification' drives led to the dis-placement of even more poor people. The draconian Bombay Prevention of Begging Act was invoked to justify the removal of hundreds of small vendors, hawkers, and panhandlers from the streets of Delhi (Uppal 2009 provides a good overview of such changes).

Land speculation and appropriation for (private) infrastructural develop-ment was another feature of the Asian Games that prefigured the 2010 CWG. In preparation for the Asian Games, the city received new sports stadiums, air-conditioned shopping malls, widened roads, tiered flyovers, a sprawling exhibition complex (Pragati Maidan), an amusement park, a number of luxury hotels (the Maurya Sheraton and Taj Palace), and colour television. Unsurprisingly, it was the more affluent areas of the city in central and south Delhi that received the best upgrades. As Uppal points out, while 'broader roads and flyovers were welcome additions to the city's crumbling infrastructure,' they were:

biased in their location as they were all built with the purpose of serv-
ing the '82 Games, meaning that they were not located in the most
congested places which needed them most but on routes which the
athletes would take between stadiums and the games village.

(Uppal 2009, p. 17)

Notably, these forms of appropriation held (and continue to hold) significant environmental consequences. For example, the athletes' compound
built for the Asian Games was converted into blocks of private apartments after the Games. Today, the Asian Games Village stands out as one
of the most coveted gated residential communities of south Delhi. The
actual construction of the Village, like many forms of land speculation
and appropriation infrastructural developments spurred by the hosting of
the Games, also had environmental impacts, as illustrated through the
appropriation of large parts of the Siri Fort forest – a green region in
south Delhi – for the construction project (Baviskar 2014).

It is striking that many of the changes associated with the 1982
Asian Games are good illustrations of the concepts of 'accumulation
by dispossession' and 'neoliberal urbanism' described earlier in this
chapter – even though they occurred prior to the ascent of neoliberalism in India. In the late 1970s and very early 1980s, when the event
was being planned, the Nehruvian model was still intact, and private
corporate interests – which are considered the motor of neoliberalism – played a relatively limited role in the Indian economy. How,
then, might an analysis of the 1982 Asian Games be instructive for
the contemporary context? Ravi Sundaram, a Delhi-based postcolonial scholar, offers some clues.

Sundaram points out that the demand for a sanitized, 'modern'
Delhi stretches back, at the very least, to the Masterplan of 1962,
which was based on "a vision of order, the legal separation of work,
commerce and industry, and proper civic citizenship" (2009, p. 34).
Using zoning as the chief instrument, the idea was to turn Delhi into
a city free not only from the wretchedness of rural poverty, but also
the 'traditionalism' of rural society (p. 34.). As discussed below, such
efforts to 'cleanse', 'sanitize', and 'modernize' the city have often been
underpinned by sentiments of 'bourgeois environmentalism,' where
middle class groups advocate for the creation of segregated, cleansed,
green living spaces for themselves, justified through discourses of
poor peoples as harbingers of disease and poor neighbourhoods as
sources of pollution (Baviskar 2003). The aims of the Masterplan,
however, remained largely unrealized, and through the decades that
followed, there were widespread deviations from the document. Sundaram writes:

The new postcolonial urban [made the] classic urban management
models irrelevant or simply inoperative [...]. Home workshops, markets, hawkers, small factories, small and large settlements of the working poor now spread all over the planned metropolis [...]. Productive,
non-legal proliferation has emerged as a defining component of the
new urban crisis in India.

(2009, p. 68)

Simply put, the city of people's lived experience could not be contained
by the plan.

Sundaram points to three distinct moments in which the city's elites
and middle classes, worried by the chaotic expansion of the city,
attempted to revive the Masterplan. The first was the 21-month period of
internal 'Emergency' (stretching between 1975 and 1977), when Prime
Minister Indira Gandhi suspended elections and civil liberties, and ruled
the country by executive decree. Among the well-documented excesses of
this authoritarian period were the slum clearance campaigns of Jagmohan
Malhotra, the head of the Delhi Development Authority. During this
period, as Sundaram notes, "the power of the urban government
expanded to its maximum possible," and the working poor were uprooted
from their homes and relocated to the peripheries of the city (2009,
p. 75). When the Emergency ended, however, a weak alliance of anti-
Congress parties was brought to power at the centre, and the momentum
of Malhotra's aggressive slum clearance drives could not be sustained.
With the Asian Games of 1982, Sundaram suggests, came a second
attempt to resuscitate the objectives behind the 1962 Masterplan.
A strong Congress government, led by Indira Gandhi, was back in power,
with Malhotra now positioned as the Lieutenant Governor of Delhi. The
Asian Games, in turn, paved the way for the third and most aggressive
phase – onwards from the 1990s – of aligning the growth of Delhi with
the 1962 Masterplan. While guided by roughly the same, broad vision of
creating an orderly and modern national capital city, the efforts of the
1990s were unlike those of previous phases because they coincided with
two developments already discussed: the liberalization of the economy in
1991 and the rise of a politically assertive 'new middle class.'

What Sundaram's study renders clear, however, is that the difference
between this third phase and those preceding it was a *matter of intensifica-
tion*, not the creation of something new. In other words, the model of
urban development preferred by the Indian government was one that pre-
dated the advent of neoliberalism, although neoliberalism accelerated the
thrust of change recommended by this model. Indeed, it appears that
there is an element of continuity in India's development trajectory that
undercuts the more widely accepted distinction between its 'state-led' and
'market-led' (or 'neoliberal') phases. But if there is, in fact, an anchoring

principle that goes beyond the usual contrasts between market versus state, what is its essential element? In his memoirs, Nehru's Chief Economic Advisor, I.G. Patel, provides some insight:

> For economists including myself, who grew up during the days of the nationalist struggle all notions of catching up fast with the West and developing a diversified industrial structure had an almost primordial appeal ... we swallowed easily notions such as the Big Push and 'anything you can do we can do better.' Many economists of repute had fuelled this ambition. After all, it was Arthur Lewis who had argued that India had a comparative advantage in the production of steel and should aim at producing 10 million tonnes of steel over the short-run.
>
> (1998, pp. 70–71)

Patel's remarks drive home an important point. While a lively debate has raged on within Indian policy and intellectual circuits about the benefits of states versus markets, there has been comparatively little reflection on the normative content of 'development.' In other words, if 'development' is taken to mean desirable social and economic change (see Culp 2013 for a concise exploration of the term), the fundamental question of what is 'desirable' in the first place has been comparatively less contentious than more policy-focused concerns, such as which mix of state and market in India's 'mixed economy' is best for realizing development objectives.

It is also worth noting here how such understandings of development – codified within Delhi's Masterplans – were not only harmful for the city's poorer residents, but for the environment. Such impacts were manifest in policies that resulted in 'haphazard and unplanned growth' that were ultimately unsustainable (Truelove 2011). Indeed, the First Masterplan, though seeking to build a 'hygienic' and 'properly ordered' city, resulted in the emergence of that of which it wished to eradicate – a parallel and unplanned city, home to the informal settlements of migrant workers. This parallel city was key to the rapid construction projects of the 1970s, including the constriction of the facilities for the 1982 Asian Games (Truelove 2011). Yet, Delhi's development continued with subsequent Masterplans that have consistently called for the creation of modern rationalized spaces, thus requiring and justifying the demolition of informal settlements. The resultant accumulation by dispossession and criminalization of the poor through 'exclusionary citizenship rights' have had profound environmental consequences, including restricted access to, and quality of, local resources such as clean drinking water (Truelove 2011, p. 147). As Baviskar (2014, p. 138) notes, such construction projects, including those of the 1982 Asian Games 'and their crumbling concrete traces are now added to the 2010 Commonwealth Games "legacies" – the financial, social, and ecological hangover.' It is these hangovers that are explored in the next section.

Neoliberal urbanism and the 2010 Commonwealth Games: making the connection

Concepts such as 'accumulation by dispossession' and 'neoliberal urbanization' are useful not only in making sense of the broad transformations underway in Indian cities, but also in contextualizing many of the specific changes brought upon Delhi by the 2010 Commonwealth Games.

A salient criticism of the CWG was that they harmed the city's poor and marginalized populations (see Housing and Land Rights Network, Habitat International Coalition 2010, 2011). In 2004, the Delhi government evicted more than 35,000 low-income families from a 100-acre strip of land on the banks of the River Yamuna in order to build a 'promenade' for the event (Menon-Sen 2010, p. 678). These families had lived on that land – also known as the Yamuna Pushta (*pushta* means embankment) – for decades. Some were workers who had come to Delhi during the construction boom precipitated by the Asian Games of 1982. Menon-Sen provides an impassioned account of the Yamuna Pushta evictions:

> They had not even removed their belongings and possessions from their homes when the demolition started – in many cases, children and old people were still inside when the bulldozers started rolling. The police cordoned off the area, and those who resisted were mercilessly beaten and chased away. The ruined colony was then set on fire. There were at least five deaths.
>
> (2010, p. 679)

Menon-Sen indicates that only 6,000 of the 35,000 families were resettled, and that resettlement colonies were characterized by 'miserable conditions' and located in 'the furthest outskirts of the city' (p. 679). In fact, through a series of stunning Google Earth photographs, Bhan (2009) shows that the Yamuna Pushta settlements (described by Menon-Sen 2010) have almost vanished from the face of the planet. This lends substance to Dupont's observation that 'the preparations for the CWG' were 'emblematic of the socio-spatial restructuring and makeover that have changed Delhi and affected those living on its margins over the last dozen years' (2013, p. 2).

Another criticism of the CWG, as with the 1982 Asian Games, was that it precipitated the city to adopt a zero-tolerance policy towards street vendors, hawkers, and 'beggars' (see Pandey 2010). In the run-up to the event, such marginal populations were forced off the streets and shunted to the outskirts of the city, for fear that they would give tourists a bad impression of the country. A night shelter for homeless people was demolished in the dead of Delhi's harsh winter for the same reason (Dupont 2011). During the actual event, moreover, poor people lost access to the city almost entirely. Delhi was transformed into a veritable fortress, with tens of thousands of police and paramilitary troops deployed on its streets. New security measures included the building of a 14 foot-tall fence around Nehru stadium and the installation of closed-circuit television in the main arteries of the city. Slums and open drains, which tend to run side by side in Delhi, were hidden from public view by plastic banners and tiled walls (see Associated Press 2010), while the leafy, upscale neighbourhoods at the centre of the city received fresh lighting, new pavements, and enhanced police patrolling.

The sprawling athletes' compound created for the CWG was also targeted by critics. The Commonwealth Games Village was developed by the Delhi government, in a public-private partnership with a Dubai-based real estate firm, with the intention of later turning its units into private apartments. The project was besieged by controversy from the outset, not only because of the government's lack of transparency about the deal struck with its private partner, but also because the Village was positioned on an ecologically sensitive strip of land on the floodplains of the River Yamuna. In her analysis of the building of the Games Village, Gill (2014) suggests that various state institutions were complicit in the land-grab. She reviews the complex administrative and legal procedures that would normally be required for a project of such scale to proceed, and argues that the judiciary, which waved it through, 'failed to apply its own environmental jurisprudence' (2014, p. 69).

The environmental legacy of the Commonwealth Games Village was also a concern largely ignored by government, private entities, and the media. Indeed, the precedence given to profit-making and national prestige was perhaps no better exemplified than through the decision to build the Village on the floodplain of the River Yamuna. The construction of the permanent Village led to yet one more concreted area to affect and restrict the natural functions of the floodplain to replenish ground water sources (Baviskar 2014). Once on the outskirts of the city, economic liberalization since the 1970s had spurred continued expansion eastward, rendering the floodplain too valuable to be left to farmers and wildlife. Yet, despite the protests of local farmers and environmentalists, who had suggested other sites closer to event venues for the Games Village, and of scientific experts, who argued that no new civic structures should be built on the floodplain as it impairs the water recharge functions of the floodplain ecosystem, thereby creating problems for present and future generations, construction of the Village continued in the interest of real-estate development. As Baviskar (2014) argues, the Games Village was a key component of a long-term strategy to accumulate public land for private means, dispossessing land from poorer farmers and non-human inhabitants. Ultimately, such neoliberal manoeuvring challenged discourses of sustainability attached to the CWG. Indeed, at the time, the CWG were promoted as the 'greenest ever,' and would 'leave a legacy beyond road and infrastructure; one of greater consciousness of the environment' (Gill 2014, p. 73). Despite subscribing to an 'ecological code' that promised environmentally friendly approaches to waste management, energy usage, and re-forestation, such environmental promises did not materialize in any meaningful way. Indeed, expectations of sustainability – both social and environmental – were abandoned for neoliberal self-interest (Gill 2014).

After the CWG ended, and in parallel to the 1982 Asian Games, the Village was turned into luxury apartments that now house local elites. Today, ordinary citizens enjoy only limited access to the facilities created for the CWG, even though they were built on public land and with public money. As for the multimillion-dollar sports venues, admission is possible only through buying tickets, memberships, and clearing security checks. Indeed, Boria Majumdar, a well-known Indian sports writer, has described the CWG's stadiums as 'white elephants' because of how infrequently they have been used since the event's end (2011, p. 128). Most tragically, it appears that the sports facilities built for the CWG have done little to advance the professional lives of Indian athletes. This is noted by Majumdar (2011), and also in a BBC video segment in which several athletes, including national pole vault champion Surekha Ranjit, complain about their surprisingly limited access to CWG stadiums. A Sports Authority of India official who is also interviewed in the segment responds to such complaints by arguing that 'these facilities are of "world class" standards and not for everyday use' (BBC 2014).

The displacing of low-income communities from their homes and sources of livelihood in order to make way for privatized and securitized buildings and compounds may be viewed as an instance of Harvey's notion (2004) of 'accumulation by dispossession.' Some of the mechanisms of neoliberal urbanism, as identified by Peck, Theodore, and Brenner (2009, p. 61), also line up well with many of the changes associated with the CWG. These include intensified surveillance of public spaces, destruction of low-income neighbourhoods to make way for speculative redevelopment, and construction of gated communities and urban enclaves, contributing to socio-economic and environmental unsustainability. In this regard, Follmann (2015) notes that the re-imagining of Delhi's riverfront through mega-projects – including sport mega-events – resulted in the passing of 'zones of exception' that advanced the rhetoric of environmental sustainability while bypassing statutory regulations in the interest of an elitist class. Such mega-projects, Follmann (2015, p. 2020) continues, "4are not the results of long-term planned development, but rather ad hoc reactions of the government agencies to realize long-desired iconic flag-ships on their way to become world class."

Discussion and conclusion

In India, the concept of 'development' has been consistently associated with 'catching up' with the West as rapidly as possible – moving, as it were, from the 'backward' and 'primitive' to the 'advanced' and 'modern' – and on this point there has been considerable agreement between the political left and political right of the country. Indeed, post-colonial scholars such as Nandy (1983) and Chatterjee (2004) identified competitive state nationalism, rather than liberalism or socialism, as the indomitable force behind India's model of top-down, 'catch-up' develop-ment. According to this post-colonial critique, there are numerous prob-lems with such a model, namely that it glorifies the conquest of nature (in ways associated with 'modernity'), and that it is fundamentally non-participatory. Policymaking is entrusted to only a few, often drawn from the Western-educated elite, who are seen to have the requisite background knowledge or 'expertise' to solve complex problems, and there is little effort to connect development 'solutions' to the aspirations and lived experiences of so-called ordinary people. In fact, if anything, aspirations to connect development to the masses are regarded with fear and disdain (for a trenchant critique of India's development model along these lines, see Kaviraj 2010).

The propensity to privilege the values and vision of Western-oriented elite 'experts' over those of average citizens is not, by any means, a problem unique to India. Post-colonial scholarship has been very helpful in clarifying the complex state–society relationships that lie beneath the façade of formal democracy in developing countries more generally, and in illuminating the enormous gulf that tends to exist in these countries between elite aspirations and the gritty, everyday realities experienced by the masses (see, for example, Ferguson 1994; Mitchell 2002). Crucially, this chapter shows that such com-petitive nationalism also has environmental impacts, where notions of 'bour-geois environmentalism' view and implement the empowerment of the upper and middle classes at the expense of rural and poor populations, informed through "anxieties around the breakdown of urban infrastructures […] appre-hensions about the scarcity of water and electricity, the increase in crime and disease, and the proliferation of unruly places" (Baviskar 2003, p. 96). Such visions and policies are compatible with sports events and ultimately contrib-ute to greater levels of unsustainability. As Rao (2019) notes, as affluent mem-bers of Indian society mobilize against the "polluting presence of poor people, they conveniently hide their own extractive relations with the envir-onment," and do so despite the fact that while "slum dwellers often consume minimal resources of land, water, and electricity, middle-class citizens are likely to expect continuous uninterrupted access to much larger quantities of these scarce commodities" (see also Baviskar 2011). A key lesson to be learned here is that the inclination to run roughshod over what average citizens want from 'development,' or desirable change, is not a feature unique to neoliberalism, but one that has deeper, more profound roots in the nature of (outward-looking and competitive) nationalism, particularly as it emerged in post-colonial states. The negative effects of this form of nationalism on both development and democracy have been profound.

In sum, the top-down, growth-focused orientation of both the 1982 Asian Games and the 2010 Commonwealth Games – along with the strenuous attempts by organizers to deny and erase signs of poverty and indignity in the city – are best explained by reference to the type of outward-oriented competitive nationalism described in this chapter, rather than to neoliberalism alone. The adverse environmental consequences of both events, produced by the desire to change the natural environment into prestige projects such as giant stadiums and sprawling apartment blocks, are also better explained when the vision and politics of competitive nationalism are taken into account.

Notes

1 According to Bhagwati and Desai, development economists such as Oskar Lange, Jan Tinbergen, and Paul Rosenstein-Rodan were extensively consulted in the process of developing Nehru's Second Five Year Plan, which formed the basis of India's state-led economic strategy (1970, p. 112).
2 Joshi and Little indicate that even in June 1991 – that is, one month prior to the initiation of this new economic policy – India was the 'most autarkic and heavily regulated non-communist country in the world' (1996, p. 10).
3 For more research with similar conclusions, see Ramanathan (2006), Follmann (2015), and Banerjee-Guha (2009).

References

Associated Press, 2010. Delhi Hides Workers and Beggars as Games Near. *New York Times*, September 28. www.nytimes.com/2010/09/29/sports/29iht-GAMES.html
Banerjee-Guha, S., 2009. Neoliberalising the 'Urban': New Geographies of Power and Injustice in Indian Cities. *Economic and Political Weekly*, 44(22), 95–107.
Baviskar, A., 2003. Between Violence and Desire. Space, Power, and Identity in the Making of Metropolitan Delhi. *International Social Sciences Journal*, 175, 89–98.
Baviskar, A., 2011. Cows, Cars and Cycle-rickshaws: Bourgeois Environmentalists and the Battle for Delhi's Streets. In: A. Baviskar and R. Ray, eds. *Elite and Everyman*. London: Routledge, 391–418.
Baviskar, A., 2014. Dreaming Big: Spectacular Events and the 'World-class' City: The Commonwealth Games in Delhi. In: J. Grix, ed. *Leveraging Legacies from Sports Mega-Events: Concepts and Cases*. Basingstoke, UK: Palgrave Macmillan, 130–141.
Baviskar, A., and Ray, R., eds., 2011. *Elite and Everyman: The Cultural Politics of the Indian Middle Classes*. New Delhi, India: Routledge.

BBC, 2014. Delhi's Commonwealth Legacy, July 14. www.bbc.com/news/business-28397191

Bénit-Gbaffou, C., 2008. In the Shadow of 2010: A Fast-tracked Local Democracy, or How to Get Rid of the Poor in Ellis Park Development Project, Johannesburg. In: U. Pillay, R. Tomlinson, and O. Brass, eds. *Development and Dreams: Urban Development Implications of 2010 Soccer World Cup*. Pretoria, South Africa: HSRC Press, 200–222.

Bhagwati, J.N., and Desai, P., 1970. *India: Planning for Industrialization*. Oxford, UK: Oxford University Press.

Bhan, G., 2009. 'This is No Longer the City I Once Knew': Evictions, the Urban Poor and the Right to the City in Millennial Delhi. *Environment and Urbanization*, 21(1), 127–142.

Black, D., and Northam, K., 2016. Mega-events and 'Bottom-up' Development: Beyond Window Dressing? In: G. Anderson and C. Kukucha, eds. *International Political Economy*. Oxford, UK: Oxford University Press, 436–452.

Boykoff, J., and Mascarenhas, G., 2016. Rio 2016: Urban Policies and Environmental Impacts. *Idées d'Amériques*, Spring–Summer, 2–5.

Brosius, C., 2010. *India's Middle Class: New Forms of Urban Leisure, Consumption and Prosperity*. New York: Routledge.

Chakravarty, S., 1987. *Development Planning: The Indian Experience*. Oxford, UK: Clarendon Press.

Chandrashekhar, C.P., and Ghosh, J., 2002. *The Market that Failed: A Decade of Neoliberal Reform in India*. New Delhi, India: Leftword.

Chatterjee, P., 2004. *The Politics of the Governed*. New York: Columbia University Press.

Cornelissen, S., 2011. More Than a Sporting Chance? Appraising the Sport for Development Legacy of the 2010 FIFA World Cup. *Third World Quarterly*, 32(3), 503–529.

Culp, J., 2013. Development. In: D. Moellendorf and H. Widdows, eds. *The Handbook of Global Ethics*. Abingdon, UK: Routledge, 170–181.

Darnell, S., and Millington, R., 2015. Modernization, Neoliberalism, and Sports Mega-events: Evolving Discourses in Latin America. In: R. Gruneau and J. Horne, eds. *Mega-events and Globalization: Capital and Spectacle in a Changing World Order*. Abingdon, UK: Routledge, 65–80.

Death, C., 2011. 'Greening' the 2010 FIFA World Cup: Environmental Sustainability and the Mega-event in South Africa. *Journal of Environmental Policy & Planning*, 13(2), 99–117.

Dupont, V., 2011. The Dream of Delhi as a Global City. *International Journal of Urban and Regional Research*, 35(3), 533–554.

Dupont, V., 2013. Which Place for the Homeless in Delhi? Scrutiny of a Mobilisation Campaign in the 2010 Commonwealth Games Context. *South Asia Multidisciplinary Academic Journal*, 8, 2–18.

Ferguson, J., 1994. *The Anti-politics Machine: Development, Depoliticization, and Bureaucratic Power in Lesotho*. Minneapolis, MN: University of Minnesota Press.

Follmann, A., 2015. Urban Mega-projects for a 'World-class' Riverfront – the Interplay of Informality, Flexibility and Exceptionality Along the Yamuna in Delhi, India. *Habitat International*, 45(3), 213–222.

Gerschenkeron, A., 1962. *Economic Backwardness in Historical Perspective*. Cambridge, MA: Harvard University Press.

Gill, G.N., 2014. Environmental Protection and Developmental Interests: A Case Study of the River Yamuna and the Commonwealth Games, Delhi, 2010. *International Journal of Law in the Built Environment*, 6(1/2), 69–90.

Goldman, M., 2011. Speculative Urbanism and the Making of the Next World City: Speculative Urbanism in Bangalore. *International Journal of Urban and Regional Research*, 35(3), 555–581.

Harvey, D., 2004. The 'New' Imperialism: Accumulation by Dispossession. *Socialist Register*, 40, 63–87.

Harvey, D., 2006. *A Brief History of Neoliberalism*. Oxford, UK: Oxford University Press.

Housing and Land Rights Network, Habitat International Coalition, 2010. *The 2010 Commonwealth Games: Whose Wealth? Whose Commons?* New Delhi, India: Housing and Land Rights Network, Habitat International Coalition – South Asia Regional Programme.

Housing and Land Rights Network, Habitat International Coalition, 2011. *Planned Dispossession: Forced Evictions and the 2010 Commonwealth Games*. New Delhi, India: Housing and Land Rights Network, Habitat International Coalition – South Asia Regional Programme.

Joshi, V., and Little, I.M.D., 1996. *India's Economic Reforms, 1991–2001*. Oxford, UK: Oxford University Press.

Kaviraj, S., 2010. *The Imaginary Institution of India: Politics and Ideas*. New York: Columbia University Press.

Khilnani, S., 1997. *The Idea of India*. New York: Farrar, Strauss & Giroux.

Kohli, A., 2004. *State-directed Development: Political Power and Industrialization in the Global Periphery*. Princeton, NJ: Princeton University Press.

Law, A., and Mooney, G., 2012. Competitive Nationalism: State, Class, and the Forms of Capital in Devolved Scotland. *Environment and Planning C: Government and Policy*, 30, 62–77.

Majumdar, B., 2011. Commonwealth Games 2010: The Index of a 'New' India? *Social Research*, 78(1), 231–254.

Matheson, V.A., and Baade, R.A., 2004. Mega-sporting Events in Developing Nations: Playing the Way to Prosperity? *South African Journal of Economics*, 72(5), 1085–1096.

Mathur, H.M., 2013. *Displacement and Resettlement in India: The Human Cost of Development*. Abingdon, UK: Routledge.

Menon-Sen, K., 2010. Delhi and CWG 2010: The Games behind the Games. *Journal of Asian Studies*, 69(3), 677–681.

Mitchell, T., 2002. *Rule of Experts: Egypt, Technopolitics, Modernity*. Berkeley, CA: University of California Press.

Myrdal, G., 1957. *Economic Theory and Underdeveloped Regions*. New York: Harper Torch Books.

Nandy, A., 1983. *The Intimate Enemy: Loss and Recovery of Self Under Colonialism*. New Delhi, India: Oxford University Press.

Nayar, B.R., 2001. *Globalization and Nationalism: The Changing Balance in India's Economic Policy, 1950–2000*. New Delhi, India: SAGE Publications.

Nurske, E., 1953. *Problems of Capital Formation in Developing Countries*. Oxford, UK: Blackwell.

Pandey, G., 2010. Delhi Street Vendors Evicted before Commonwealth Games. BBC South Asia, August 20. www.bbc.com/news/world-south-asia-10716139

Patel, I.G., 1998. *Glimpses of Indian Economic Policy: An Insider's View*. Oxford, UK: Oxford University Press.

Peck, J., Theodore, N., and Brenner, N., 2009. Neoliberal Urbanism: Models, Moments, Mutations. *SAIS Review*, XXIX(1), 49–66.

Pillay, U., and Bass, O., 2008. Mega-events as a Response to Poverty Reduction: The 2010 FIFA World Cup and Its Urban Development Implications. *Urban Forum*, 19, 329–346.

Ramanathan, U., 2006. Illegality and the Urban Poor. *Economic and Political Weekly*, 41(29), 3193–3197.

Rao, U., 2019. Comment on Baviskar, Amita (2003), Between Violence and Desire. Space, Power, and Identity in the Making of Metropolitan Delhi. *International Social Science Journal*, 175, 89–98.

Rosenstein-Rodan, P., 1943. Problems of Industrialization of Eastern and South-Eastern Europe. *Economic Journal*, 53, 210–211.

Silk, M., 2014. Neoliberalism and Sports Mega-events. In: J. Grix, ed. *Leveraging Legacies from Sports Mega-events: Concepts and Cases*. London: Palgrave Pivot, 50–60.

Sundaram, R., 2009. *Pirate Modernity: Delhi's Media Urbanism*. New York: Routledge.

Truelove, Y., 2011. (Re-)conceptualizing Water Inequality in Delhi, India through a Feminist Political Ecology Framework. *Geoforum*, 42(2), 143–152.

United Nations, 2016. *Sport and the Sustainable Development Goals: An Overview Outlining the Contribution of Sport to the SDGs*. New York: United Nations.

Uppal, V., 2009. The Impact of the Commonwealth Games 2010 on Urban Development of Delhi. *Theoretical and Empirical Researches in Urban Management*, 4(10), 7–29.

Wyatt, A., 2005. Building the Temples of Postmodern India: Economic Constructions of National Identity. *Contemporary South Asia*, 14(4), 465–480.

Zirin, D., 2014. *Brazil's Dance with the Devil: The World Cup, the Olympics, and the Struggle for Democracy*. Chicago, IL: Haymarket Books.

Chapter 10

Examining outdoor recreation as an approach to promote youth engagement in environmental activism

Tanya Halsall and Tanya Forneris

Introduction

Recognizing that youth will play a critical role in shaping our future (Patton et al., 2016) and taking consideration of the environmental crisis at hand, it is imperative to examine new ways to engage young people in environmental efforts so that they can contribute to viable solutions related to environmental sustainability. This chapter examines Sport for Development (SfD) opportunities, with a specific focus on outdoor recreation programming as a significant opportunity to promote youth connection to nature and engagement in environmental sustainability in order to enhance positive environmental outcomes in the future. It highlights the health benefits that can be derived from engaging in nature and emphasizes that this may help to moderate human investment in the environment and influence engagement in sustainable behaviors. This chapter also presents a range of program approaches and applies ecological systems theory (EST) to examine how programming can deliver benefits, both for youth as well as for the community and environment. Finally, it reviews the application of EST within an Indigenous context and discusses the related concept of two-eyed seeing.

An overview of outdoor recreation programming

The Sustainable Development Goals (SDGs) are a set of interrelated economic, social and environmental objectives that were adopted by world leaders in 2015 to guide global development leading up to the year 2030 (United Nations, 2015). The United Nations Office on Sport for Development and Peace (2017) recognized the contributions that SfD programming can provide in achieving the SDGs, including those directly focused on the environment, notably SDGs 6 (clean water and sanitation), 12 (responsible consumption and production), 13 (climate action), 14 (life below water) and 15 (life on land). Furthermore, environmental education has been outlined as a valuable area of action within sport in order to promote sustainable development (Jagemann, 2004) and researchers have also recognized that youth have an important role to play regarding environmental conservation as they will be increasingly impacted by environmental issues and climate change (Sayal et al., 2016).

To date, SfD programming has primarily focused on using sport as a vehicle to foster positive outcomes for youth, including life skill development, enhanced education, improved health and the reduction of conflict (Kidd, 2008). A few studies have examined the influence that sport, sport organizations, facilities and mega-events may have on environmental issues (Casper, Pfahl, & McSherry, 2012; Jin, Zhang, Ma, & Connaughton, 2011; Mallen & Chard, 2012). However, the main focus has been on how to reduce negative impacts, rather than examining how SfD programming can be used to enhance environmental sustainability. Sport-based positive youth development (PYD) programming is a promising SfD practice that may contribute key insights to sport interventions for the environment. PYD interventions focus on the promotion of positive developmental assets and influences that lead to successful development (Roth, Brooks-Gunn, Murray, & Foster, 1998) and recognize that interventions must go beyond resolving youth problems in order to prepare them to successfully navigate life's challenges (Pittman, 2001; Pittman, Irby, Tolman, Yohalem, & Ferber, 2003). Youth leadership has been identified as one of the critical components of effective PYD programs (Lerner et al., 2006).

Although the majority of research in PYD has examined the impact of more traditional sport-based programs, there has been an expanding body of research in more recent years that examines outdoor recreation as a particular context in which positive developmental outcomes can be supported (Browne, Garst, & Bialeschki, 2011; Ferrari & McNeely, 2007; Halsall, Kendellen, Bean, & Forneris, 2016; Reis, Ng-A-Fook, & Glithero, 2015). Outdoor PYD programming may be unique compared to other sport-based interventions as it includes the added benefit of exposing participants to the natural environment and may support enhanced well-being and the development of leadership within ecological conservation.

There is a range of outdoor recreation programming targeted at children and youth that has been found to promote positive health and development outcomes. One promising practice within outdoor recreation is the Forest School model. This outdoor education model was originally developed in Scandinavia in the late 1950s (MacEachren, 2013; Slade, Lowery, & Bland, 2013) and involves experiential learning within a natural space (MacEachren, 2013; Ridgers, Knowles, & Sayers, 2012; Slade et al., 2013). In Canada, the Forest School model is also informed by Indigenous land-based pedagogy that emphasizes self-directed learning and includes the integration of story (Forest School Canada, 2014). Researchers suggest that learning in an outdoor environment, such as within a Forest School, can benefit children and youth through the improvement of academic outcomes (Williams & Dixon, 2013) and increased engagement in physical activity (Mygind, 2007).

Outdoor adventure programming is another example of nature-based programs that can support positive developmental outcomes for youth. Often referred to as adventure therapy, these programs comprise interventions that involve natural settings to promote positive change that incorporate an element of challenge, participant engagement, meaningful learning through natural consequences and the inclusion of strong therapeutic support (Gass, Gillis, & Russell, 2012). Although this definition places a focus on the medical model and clinical populations, in Canada the term "adventure therapy" is not widely used and many interventions take a more holistic perspective to promote general well-being (Ritchie, Patrick, Corbould, Harper, & Oddson, 2016). There are a range of formats that include day-only programs, multiple-day expeditions, center-based programs, summer camps and journey-based programs (Gass et al., 2012).

Residential camp is a programming model whereby youth participate in an extended stay (American Camping Association, 1998) and experience community living within an outdoor recreational setting (American Camping Association, 2006). Residential camps have also been found to create positive impacts on children and youth (Bialeschki, Henderson, & James, 2007; Garst & Johnson, 2005; Henderson et al., 2007; Thurber, Scanlin, Scheuler, & Henderson, 2007). Researchers have identified a range of positive youth outcomes that can be developed through camp experiences, including enhanced independence, spirituality, self-esteem, hopefulness, positive identity, social skills and levels of responsibility (Bialeschki et al., 2007; Garst & Johnson, 2005; Halsall et al., 2016; Henderson et al., 2007; Thurber et al., 2007). Other research has also reported on the benefits of camp that were related to engagement in an outdoor context, including increased awareness of the environment and appreciation of nature (DeGraaf & Glover, 2003).

There is also an expanding body of research that examines land-based programming, an Indigenous outdoor recreation practice informed by Indigenous education. Indigenous education reflects the importance of land in Indigenous culture and the intrinsic personal connections to community and the environment. Cajete (1994) suggests that Indigenous education "unfolds within an authentic context of community in Nature" (p. 30) and emphasizes that "A sacred view of Nature permeates its foundational process of teaching and learning" (p. 29).

There are several examples of land-based education programming for youth in the literature. Ritchie and colleagues (2014) examined an outdoor adventure-based youth leadership program that was implemented with youth from the Wikwemikong community on Manitoulin Island in Ontario. This was an expedition-based program that involved a ten-day canoe trip that included cultural activities including ceremonies, traditional teachings from Elders and explorations of hunting grounds and local medicines. They found that youth participants' resilience increased after participation in the program (Ritchie et al., 2014). In Moose Cree First Nation, Gaudet (2016) studied two land-based programs for youth: Project George and Milo Pimatisiwin. Activities included fishing and trapping expeditions, sweat lodge ceremonies and traditional teachings related to hunting, gathering and preparing food and community feasts. They also recorded and broadcast a segment on the local radio that described the seven *pimatisiwin* teachings modules. These were later translated to Cree and shared on the community app. Alfred (2014) describes the Akwesasne Cultural Restoration, a land-based apprenticeship program that was implemented in Akwesasne, a Mohawk community near Cornwall whose territory straddles Ontario, Quebec and New York State. This community-driven program was designed to enhance youth cultural knowledge and experience in order to promote land-based practices such as fishing, hunting, horticultural practices and use of traditional medicines and basket-making.

Environmental issues and climate change are being felt more intensely by Indigenous populations and this is having a profound impact on their health and well-being (Harada et al., 2005; McQuigge, 2012; Sarkar, Hanrahan, & Hudson, 2015). For example, in examining a Sport for Development leadership program for Indigenous youth, researchers found that many communities participating in the program across Ontario were coping with critical environmental issues within their communities, including toxicity of local bodies of water, exposure to petrochemical pollution and floods (Halsall, 2016). In research with Inuit communities in Nunatsiavut, Labrador, Willox and colleagues (2013) explored the impact of climate change on health and well-being. They found that changes in temperature, weather patterns and ice stability were creating barriers for community members' participation in land-based practices and this was resulting in diminished cultural identity and well-being. These findings are very concerning considering that Indigenous health and well-being is intrinsically connected to the land (Dell et al., 2011; Kirmayer, Tait, & Simpson, 2009).

In sum, there is strong support for the health and wellness benefits of outdoor recreation programming for children and youth. Next, we explore how these interventions may contribute to youth leadership in environmental activism and hence contribute to the reciprocal relationships posited by ecological systems theory.

Possible mechanisms of influence between outdoor recreation and environmental engagement

We argue that affinity for the environment and investment in environmental sustainability can be developed through ongoing exposure to the environment within the kind of outdoor recreation experiences discussed above.

Researchers suggest that much of contemporary society suffers from disengagement with nature as a result of industrial development, urbanization and the constraints created through modern lifestyle (Capaldi, Passmore, Nisbet, Zelenski, & Dopko, 2015; Gass et al., 2012; Keniger, Gaston, Irvine, & Fuller, 2013; Louv, 2009). In addition, excessive rules and regulations designed to reduce risk of injury in children and youth, and concerns regarding perceived liabilities along with increased levels of screen time, have also been found to contribute to children and youth spending less time outdoors (ParticipACTION, 2015).

This is concerning as interactions with nature have been found to have a positive effect on physiological states, psychological well-being and cognitive functioning. For example, research has identified that experiencing nature can promote a range of benefits, including enhanced mood, cognitive function and productivity, reduced mental fatigue, stress and anxiety, and an improvement in a variety of physical health indicators (Keniger et al., 2013). Researchers have also begun to identify the reasons why the natural environment may have an impact on well-being. Major theoretical explanations suggest that attraction to nature may be influenced by our evolutionary development, through stress reduction or through the restoration of attention (Capaldi et al., 2015). Gass and colleagues (2012) provide an example describing the process of attention restoration: "Clouds, sunsets, and moving river water capture attention because they are visually and auditorily fascinating, but in a way that does not require direct attention" (p. 107).

There also appears to be added health benefits related with being active outdoors. In a meta-analysis of studies examining exercising in a natural environment in comparison with built environments, Bowler, Buyung-Ali, Knight, and Pullin (2010) found that green exercise was associated with reduced negative emotions and improved attention. In another systematic review comparing exercising indoors with exercising in natural environments, Thompson Coon et al. (2011) identified that green exercise was associated with increased energy, greater feelings of revitalization, positive engagement, enjoyment, decreases in tension, confusion, anger and depression, enhanced satisfaction and stronger motivation to continue the activity. In addition, researchers have found that children and youth tend to be most active when engaged in outdoor play (Perry, Ackert, Sallis, Glanz, & Saelens, 2016) and outdoor exposure has been found to increase levels of physical activity (Gray et al., 2015). In order to encourage Canadian families and services to increase children's levels of outdoor activity, ParticipACTION released a position statement in 2015 that states: "Access to active play in nature and outdoors—with its risks—is essential for healthy child development. We recommend increasing children's opportunities for self-directed play outdoors in all settings—at home, at school, in child care, the community and nature" (p. 8).

Exposure to the environment through outdoor recreation programming may accrue individual benefits as well as connection to and investment in the natural environment. Ecological systems theory helps to explain the possible mechanisms through which outdoor recreation can influence youth environmental engagement. EST conceptualizes youth development within the context of multiple levels of influence derived from environmental factors such as family, school, community and socio-cultural systems (Bronfenbrenner & Morris, 2006). EST includes four major components: (1) process, (2) person, (3) context and (4) time (Bronfenbrenner, 1995, 1999; Bronfenbrenner & Morris, 2006). The process component, also called proximal process, is the most fundamental concept within the theory and is defined as the "progressively more complex reciprocal interaction between an active, evolving biopsychological human organism and the persons, objects, and symbols in its immediate external environment" (Bronfenbrenner & Morris, 2006, p. 797). The process of youth leadership development can also be understood using the concept of proximal process. Youth leadership and contribution to community have been identified as behaviors that result from healthy youth development which enhances a mutually adaptive interaction between youth and their environment (Lerner et al., 2005). This interaction can be understood using the construct of proximal processes as representing the interaction between youth leadership skill development and associated positive outcomes within the developmental context. For example, programming that is designed to enhance youth leadership skills through life skills development can also benefit the community through the promotion of youth engagement in voluntary service and community action. In their study examining a SfD youth leadership program for Indigenous youth, Halsall and Forneris (2016c) found that the program had positive impacts on youth leadership skills as well as the community, including an enhanced sense of community and greater community partnerships such as increased volunteerism and collaboration between organizations. This mutually reciprocal interaction between youth and community may also apply in programs that promote youth engagement in environmental efforts; in this setting, contextual impacts would be exhibited as environmental benefits. For example, youth programming that focuses on developing leadership skills related to environmental policy might support youth advocacy efforts to influence decision-making with respect to environmental protection and sustainable practices.

Researchers have identified that currently many youth: (1) have relatively low levels of knowledge about the underlying mechanisms and scientific perspectives on environmental issues, (2) perceive that climate change is primarily a government responsibility and (3) have a general lack of self-efficacy with respect to environmental engagement (Corner et al., 2015). Recognizing these issues, it is very important to invest in grassroots efforts that directly engage young people with trusted leaders, apply experiential and peer-to-peer approaches that help to enhance their interest and support their confidence and ability related to climate change efforts (Corner et al., 2015). Outdoor recreation programs can promote this kind of development of youth leadership and specifically target these skills to environmental conservation. In their study, Browne et al. (2011) examined the effectiveness of Camp 2 Grow, a nature-based day and residential camp program designed to promote leadership, personal development and engagement in environmental sustainability. They found that participation in a nature-based camp enhanced independence, leadership, confidence in problem-solving and affinity for nature. In another example, Students on Ice, an expedition-based program that was initiated in 1999, provided an opportunity for university students from across the world to visit the Polar Regions (Reis et al., 2015). This program was designed to expose participants to the natural environment, increase knowledge and values regarding the environment, develop problem-solving related to environmental challenges and to promote engagement in environmental activism (Reis et al., 2015). Similar to other PYD programs that promote youth leadership and community contribution, research has identified that engagement in environmental efforts also promotes positive developmental outcomes and well-being in youth (Browne, Garst, & Bialeschki, 2011; Waite, Goodenough, Norris, & Puttick, 2016).

Indigenous approaches to environmental education

Looking forward, further research is needed to examine the reciprocal impact that youth participation in outdoor recreation may have. In addition, more work is needed to increase the opportunities for youth engagement and participation in outdoor recreation, particularly given the findings related to health outcomes. For instance, models like the Forest School have not yet been broadly adopted, yet they are designed to have population-level impacts if they are implemented within publicly funded schools. Creating opportunities for all children and youth to appreciate and benefit from time spent in nature may have population-level health and education benefits and may create a future population that is more invested in the natural environment and willing to contribute to conservation efforts. Jagemann (2004) warns against possible damage to the natural environment when there are too many sport participants impacting vulnerable contexts. He emphasizes the importance of using urban greenspace and residential areas for physical activities so that more susceptible settings are spared. This aligns with recent research and programming that applies EST to enhance participation in physical activity within urban environments and through active living (Sallis et al., 2006). These initiatives create more local opportunities within urban greenspace for children and youth to spend active time outdoors, whether within school programming or during day-to-day activities such as time spent in leisure and travel.

According to EST, in order to be effective, the reciprocal interactions between an individual and their environment must be exerted over an extended period of time (Bronfenbrenner & Morris, 2006). As such, long-term programs such as Forest School may provide added benefit over interventions that are delivered over a shorter timeframe. Conversely, programs that bring participants out of their typical context, such as outdoor adventure programs and camps, also temporarily remove the influence of the other contexts during programming (i.e. family, school and community). In cases where these typical contexts are exerting a negative influence, there may be an enhanced influence of the intervention during the time of programming. Relatedly, extended program exposure over time has been identified as being more effective in the promotion of youth development (Catalano, Berglund, Ryan, Lonczak, & Hawkins, 2002) and researchers have also identified that intensity or frequency of intervention exposure is related to a sense of belonging and engagement (Akiva, Cortina, Eccles, & Smith, 2013). Future research should examine the comparative effects between long-term programming and out-of-context intensive short-term programming and examine the influence on participant environmental engagement.

The EST approach has also been used to better understand sport programming in Indigenous communities. Two-eyed seeing is a concept that Mi'kmaq Elder Albert Marshall used to describe an approach to understanding that combines both Indigenous and non-Indigenous perspectives and emphasizes integrating strengths from each cultural viewpoint (Lavallée & Lévesque, 2012). Two-eyed seeing has been applied as a research framework within sport and physical activity research and it has also been used to adapt the EST to enhance its application within Indigenous communities (see Baillie et al., 2016; Lavallée & Lévesque, 2012). For example, Lavallée and Lévesque (2012) created a hybrid EST that incorporates the medicine wheel and integrates nature and spirit as systems of influence. This model depicts decolonizing sport and physical activity interventions delivered at multiple system levels within intrapersonal, interpersonal, organizational, community, policy/system/environment, mother earth and all of creation in order to promote holistic health and balance between the physical, mental, emotional and spiritual life aspects.

These adaptations of EST further highlight the influence of the natural environment and the potential reciprocal interaction with the individual. Since these models conceptualize nature as a specific context of influence, they may be useful to apply to future research to conceptualize proximal processes between individuals and the natural environment and examine how these processes may enhance individual well-being and promote individual engagement within environmental conservation.

The concept of two-eyed seeing can also be applied in order to enhance programming. For example, integrating Western technology into land-based approaches can make the programs more accessible to youth and more applicable to their modern lifestyle. Similar to the general youth population, mobile technology and social media has become a big part of many Indigenous youths' lives. Programs for Indigenous youth should take advantage of social media as an opportunity to engage youth and facilitate communication. Within their youth leadership program for Indigenous youth, Sport for Development organization Right to Play integrated Facebook as a way to support communication and promote the program while also emphasizing the learning of traditional cultural values and practices (Halsall & Forneris, 2016b).

There are also many successful examples of research that applies Photovoice to facilitate Indigenous youth empowerment, engagement in community development and reflection on cultural identity (Baillie et al., 2016; Halsall & Forneris, 2016a; McHugh, Coppola, & Sinclair, 2013; Pearce & Coholic, 2013; Young et al., 2013). In her research examining land-based programming for Indigenous youth, Gaudet (2016) used a community app to share traditional Cree teachings with community youth. To highlight the importance of increasing the accessibility of programming for modern Indigenous youth, she suggests that: "Youth have a unique circumstance in that they must learn who they are as Cree people in the contemporary Western world. Our youth and our people would benefit greatly from learning to walk in both worlds" (Gaudet, 2016, p. 191). Similarly, Cajete (1994) suggests that program processes must take account of the current context while building on historical developments: "we engineer the new reality built upon earlier ones, while simultaneously addressing the needs, and acting in the sun, of our times" (p. 27). These technologies can be used to enhance Indigenous youth engagement within land-based programming and can also be applied to raise awareness about the cultural importance of the land and current environmental issues.

Conversely, Indigenous practices and approaches can be used to improve western-based programming. For example, the Canadian Forest School model is informed by Indigenous land-based pedagogy that emphasizes self-directed experiential learning and the integration of stories (Forest School Canada, 2014). This approach utilizes the strengths from Indigenous education to benefit children and youth from the general population. Researchers have identified that stories can enhance learning as it enhances meaning, relevance and learner engagement and promotes retention (Davis, 2014; Green, 2004; Lawrence & Paige, 2016; Phillips, 2013). Furthermore, experiential education, a traditional Indigenous approach, can promote critical thinking and ability to apply new knowledge, a deeper understanding of content, and the ability to engage in life-long learning across contexts (Eyler, 2009). Cajete (1994) argues that the Western education system in general would benefit from a greater adoption of Indigenous philosophy and practice:

> What underlies the crisis of American education is the crisis of modern man's identity and his cosmological disconnection from the natural world... Traditional American Indian forms of education must be considered conceptual wellsprings for the "new" kinds of educational thought that can address the tremendous challenges of the 21st-century. Tribal education presents models and universal foundations to transform American education and develop a "new" paradigm for curricula that will make a difference for Life's Sake.
>
> (pp. 25–26)

Finally, another added benefit related to two-eyed seeing and combining Indigenous with Western values and approaches is that it can foster reconciliation. Acknowledging and valuing both perspectives supports the recognition of cultural strengths. Paraschak (2013) recommends applying strengths-based approaches as they promote the identification of successes, provide a holistic perspective and place an emphasis on cultural similarities, rather than differences. With respect to environmental issues, non-Indigenous individuals may develop an increased connection and respect for the environment through an improved understanding of Indigenous culture and recognition of the value of the environment within the Indigenous perspective.

Conclusion

This chapter presented the argument that outdoor recreation programs have the potential to benefit youth as well as the environment. Using outdoor recreation programming to create a personal connection to and affinity for nature, as well as the knowledge and skills needed to engage in environmental sustainability, will help to create a population of future adults who are willing and able to promote environmental sustainability. In addition, there is an opportunity for researchers to use EST to support the exploration of youth development, environmental activism and sustainable development and to help explain reciprocal influences. Finally, both researchers and practitioners need to recognize the value of integrating models of outdoor recreation and Indigenous land-based education, as such an approach can have benefits for both Indigenous and non-Indigenous youth as well as for our collective future.

References

Akiva, T., Cortina, K. S., Eccles, J. S., & Smith, C. (2013). Youth belonging and cognitive engagement in organized activities: A large-scale field study. *Journal of Applied Developmental Psychology*, 34(5), 208–218.

Alfred, T. (2014). The Akwesasne cultural restoration program: A Mohawk approach to land-based education. *Decolonization: Indigeneity, Education & Society*, 3(3), 134–144.

American Camping Association. (1998). *Accreditation standards for camp programs and services*. Martinsville, IN: American Camping Association.

American Camping Association. (2006). *Inspirations: Developmental supports and opportunities of youths' experiences at camp*. Martinsville, IN: American Camping Association.

Baillie, C. P., Johnson, A. M., Drane, S., LePage, R., Whitecrow, D., & Lévesque, L. (2016). For the community, by the community: Working with youth to understand the physical activity-environment relationship in First Nations communities. *Youth Engagement in Health Promotion*, 1(2), 1–33.

Bialeschki, M. D., Henderson, K. A., & James, P. A. (2007). Camp experiences and developmental outcomes for youth. *Child and Adolescent Psychiatric Clinics of North America*, 16(4), 769–788.

Bowler, D. E., Buyung-Ali, L. M., Knight, T. M., & Pullin, A. S. (2010). A systematic review of evidence for the added benefits to health of exposure to natural environments. *BMC Public Health*, 10(1), 456.

Bronfenbrenner, U. (1995). Developmental ecology through space and time: A future perspective. In P. Moen, G. H. Elder Jr., & K. Luscher (Eds.), *Examining lives in context: Perspective son the ecology of human development* (pp. 619–647). Washington, DC: American Psychological Association.

Bronfenbrenner, U. (1999). Environments in developmental perspective: Theoretical and operational models. In S. L. Friedman & T. D. Wachs (Eds.), *Measuring the environment across the lifespan: Emerging methods and concepts* (pp. 3–28). Washington, DC: American Psychological Association.

Bronfenbrenner, U., & Morris, P. A. (2006). The bioecological model of human development. In R. M. Lerner & W. Damon (Eds.), *Handbook of child psychology: Theoretical models of human development* (pp. 793–828). Hoboken, NJ: John Wiley.

Browne, L. P., Garst, B. A., & Bialeschki, D. (2011). Engaging youth in environmental sustainability: Impact of the Camp 2 Grow Program. *Journal of Park and Recreation Administration*, 29(3), 70–85.

Cajete, G. (1994). *Look to the mountain: An ecology of Indigenous education*. Durango, CO: Kivaki Press.

Capaldi, C. A., Passmore, H. A., Nisbet, E. K., Zelenski, J. M., & Dopko, R. L. (2015). Flourishing in nature: A review of the benefits of connecting with nature and its application as a wellbeing intervention. *International Journal of Wellbeing*, 5(4), 1–16.

Casper, J., Pfahl, M., & McSherry, M. (2012). Athletics department awareness and action regarding the environment: A study of NCAA athletics department sustainability practices. *Journal of Sport Management*, 26(1), 11–29.

Catalano, R., Berglund, M., Ryan, J., Lonczak, H., & Hawkins, J. (2002). Positive youth development in the United States: Research findings on evaluations of positive youth development programs. *Prevention and Treatment*, 5(15), 1–111.

Corner, A., Roberts, O., Chiari, S., Völler, S., Mayrhuber, E. S., Mandl, S., & Monson, K. (2015). How do young people engage with climate change? The role of knowledge, values, message framing, and trusted communicators. *Wiley Interdisciplinary Reviews: Climate Change*, 6(5), 523–534.

Davis, J. (2014). Towards a further understanding of what Indigenous people have always known: Storytelling as the basis of good pedagogy. *First Nations Perspectives: The Journal of the Manitoba First Nations Education Resource Centre*, 6(1), 83–96.

DeGraaf, D., & Glover, J. (2003). Long-term impacts of working at an organized camp for seasonal staff. *Journal of Park & Recreation Administration*, 21(1), 1–20.

Dell, C. A., Seguin, M., Hopkins, C., Tempier, R., Mehl-Madrona, L., Dell, D., ... Mosier, K. (2011). From benzos to berries: Treatment offered at an Aboriginal youth solvent abuse treatment centre relays the importance of culture. *Canadian Journal of Psychiatry*, 56(2), 75–83.

Eyler, J. (2009). The power of experiential education. *Liberal Education*, 95(4), 24–31.

Ferrari, T. M., & McNeely, N. N. (2007). Positive youth development: What's camp counseling got to do with it? Findings from a study of Ohio 4-H camp counselors. *Journal of Extension*, 45(2), 2RIB7.

Forest School Canada. (2014). *Forest and Nature School in Canada: A head, heart, hands approach to outdoor learning.* Nepean, Canada: Forest School Canada. Retrieved on July 15, 2019 from: http://childnature.ca/wp-content/uploads/2017/10/FSC-Guide-1.pdf

Garst, B. A., & Johnson, J. (2005). Adolescent leadership skill development through residential 4-H camp counseling. *Journal of Extension*, 43(5), 1–6.

Gass, M. A., Gillis, L., & Russell, K. C. (2012). *Adventure therapy: Theory, research, and practice.* Abingdon, UK: Routledge.

Gaudet, J. C. (2016). An Indigenous methodology for coming to know Milo Pimatisiwin as land-based initiatives for Indigenous youth (Doctoral dissertation, University of Ottawa).

Gray, C., Gibbons, R., Larouche, R., Sandseter, E. B. H., Bienenstock, A., Brussoni, M., ... Power, M. (2015). What is the relationship between outdoor time and physical activity, sedentary behaviour, and physical fitness in children? A systematic review. *International Journal of Environmental Research and Public Health*, 12(6), 6455–6474.

Green, M. C. (2004). Storytelling in teaching. *Association for Psychological Science Observer*, 17(4), 37–39.

Halsall, T. (2016). Evaluation of a sports-based positive youth development program for First Nations youth: Experiences of community, growth and youth engagement (Doctoral dissertation, University of Ottawa).

Halsall, T., & Forneris, T. (2016a). Behind the scenes of youth-led community events: A participatory evaluation approach using Photovoice in a Canadian First Nation community. *Youth Engagement in Health Promotion*, 1(2), 1–40.

Halsall, T., & Forneris, T. (2016b). Challenges and strategies for success of a sport-for-development programme for First Nations, Métis and Inuit youth. *Journal of Sport for Development*, 4(7), 39–57.

Halsall, T., & Forneris, T. (2016c). Evaluation of a leadership program for First Nations, Métis and Inuit youth: Stories of positive youth development and community engagement. *Applied Developmental Science*, 22(2), 1–14.

Halsall, T., Kendellen, K., Bean, C., & Forneris, T. (2016). Processes that facilitate positive youth development through residential camp: Understanding the importance of leader characteristics for developing supportive relationships and strategies for youth engagement. *Journal of Park and Recreation Administration*, 34(4), 20–35.

Harada, M., Fujino, T., Oorui, T., Nakachi, S., Nou, T., Kizaki, T., ... Ohno, H. (2005). Follow-up study of mercury pollution in Indigenous tribe reservations in Province of Ontario, Canada, 1975–2002. *Bulletin of Environmental Contamination and Toxicology*, 74(4), 689–697.

Henderson, K. A., Bialeschki, M. D., Scanlin, M. M., Thurber, C., Whitaker, L. S., & Marsh, P. E. (2007). Components of camp experiences for positive youth development. *Journal of Youth Development*, 1(3), 1–12.

Jagemann, H. (2004). Sports and the environment: Ways towards achieving the sustainable development of sport. *The Sport Journal*, 7(1).

Jin, L., Zhang, J. J., Ma, X., & Connaughton, D. P. (2011). Residents' perceptions of environmental impacts of the 2008 Beijing Green Olympic Games. *European Sport Management Quarterly*, 11(3), 275–300.

Keniger, L. E., Gaston, K. J., Irvine, K. N., & Fuller, R. A. (2013). What are the benefits of interacting with nature? *International Journal of Environmental Research and Public Health*, 10(3), 913–935.

Kidd, B. (2008). A new social movement: Sport for development and peace. *Sport in Society*, 11(4), 370–380.

Kirmayer, L. J., Tait, C. L., & Simpson, C. (2009). The mental health of Aboriginal peoples in Canada: Transformations of identity and community. In L. J. Kirmayer & G. G. Valaskis (Eds.), *Healing traditions: The mental health of Aboriginal peoples in Canada* (pp. 3–35). Vancouver, Canada: UBC Press.

Lavallée, L., & Lévesque, L. (2012). Two-eyed seeing: Physical activity, sport, and recreation promotion in Indigenous communities. In J. Forsyth & A. R. Giles (Eds.), *Aboriginal peoples and sport in Canada* (pp. 206–228). Vancouver, Canada: UBC Press.

Lawrence, R. L., & Paige, D. S. (2016). What our ancestors knew: Teaching and learning through storytelling. *New Directions for Adult and Continuing Education*, 149, 63–72.

Lerner, R. M., Lerner, J. V., Almerigi, J., Theokas, C., Phelps, E., Naudeau, S., ... von Eye, A. (2006). Toward a new vision and vocabulary about adolescence: Theoretical, empirical, and applied bases of a "positive youth development" perspective. In L. Balter & C. Tamis-Lemonda (Eds.), *Child psychology: A handbook of contemporary issues* (pp. 445–469). New York: Taylor & Francis.

Lerner, R. M., Lerner, J. V., Almerigi, J. B., Theokas, C., Phelps, E., Gestsdottir, S., ... Smith, L. M. (2005). Positive youth development, participation in community youth development programs, and community contributions of fifth-grade adolescents: Findings from the first wave of the 4-H study of positive youth development. *Journal of Early Adolescence*, 25(1), 17–71.

Louv, R. (2009). Do our kids have nature-deficit disorder. *Educational Leadership*, 67 (4), 24–30.

MacEachren, Z. (2013). The Canadian Forest School movement. *Learning Landscapes*, 17(1), 219–233.

Mallen, C., & Chard, C. (2012). "What could be" in Canadian sport facility environmental sustainability. *Sport Management Review*, 15(2), 230–243.

McHugh, T.-L. F., Coppola, A. M., & Sinclair, S. (2013). An exploration of the meanings of sport to urban Aboriginal youth: A Photovoice approach. *Qualitative Research in Sport, Exercise and Health*, 5(3), 291–311.

McQuigge, M. (2012). Two Ontario First Nations show signs of ongoing mercury poisoning: Report. *National Post*. Retrieved on November 14, 2012 from: http://news.nationalpost.com/2012/06/04/two-ontario-first-nations-communities-show-signs-of-ongoing-mercury-poisoning-report/

Mygind, E. (2007). A comparison between children's physical activity levels at school and learning in an outdoor environment. *Journal of Adventure Education & Outdoor Learning*, 7(2), 161–176.

Paraschak, V. (2013). Hope and strength(s) through physical activity for Canada's Aboriginal peoples. In C. Hallinan & B. Judd (Eds.), *Native games: Indigenous peoples and sports in the post-colonial world* (pp. 229–246). Bingley, UK: Emerald Group Publishing.

ParticipACTION. (2015). *The biggest risk is keeping kids indoors: The 2015 ParticipACTION report card on physical activity for children and youth*. Toronto, Canada: ParticipACTION.

Patton, G. C., Sawyer, S. M., Santelli, J. S., Ross, D. A., Afifi, R., Allen, N. B., … Kakuma, R. (2016). Our future: A Lancet commission on adolescent health and wellbeing. *The Lancet*, 387(10,036), 2423–2478.

Pearce, K., & Coholic, D. (2013). A Photovoice exploration of the lived experiences of a small group of Aboriginal adolescent girls living away from their home communities. *Pimatisiwin: A Journal of Aboriginal & Indigenous Community Health*, 11(1), 113–124.

Perry, C. K., Ackert, E., Sallis, J. F., Glanz, K., & Saelens, B. E. (2016). Places where children are active: A longitudinal examination of children's physical activity. *Preventive Medicine*, 93, 88–95.

Phillips, L. (2013). Storytelling as pedagogy. *Literacy Learning: The Middle Years*, 21 (2), ii–iv.

Pittman, K. (2001). Balancing the equation: Communities supporting youth, youth supporting communities. *Community Youth Development Journal*, 1(1), 19–24.

Pittman, K., Irby, M., Tolman, J., Yohalem, N., & Ferber, T. (2003). *Preventing problems, promoting development encouraging engagement: Competing priorities or inseparable goals?* Washington, DC: Forum for Youth Investment, Impact Strategies.

Reis, G., Ng-A-Fook, N., & Glithero, L. (2015). Provoking ecojustice—taking citizen science and youth activism beyond the school curriculum. In M. P. Mueller & D. J. Tippins (Eds.), *EcoJustice, citizen science and youth activism* (pp. 39–61). Basel, Switzerland: Springer International.

Ridgers, N. D., Knowles, Z. R., & Sayers, J. (2012). Encouraging play in the natural environment: A child-focused case study of Forest School. *Children's Geographies*, 10(1), 49–65.

Ritchie, S. D., Patrick, K., Corbould, G. M., Harper, N. J., & Oddson, B. E. (2016). An environmental scan of adventure therapy in Canada. *Journal of Experiential Education*, 39(3), 303–320.

Ritchie, S. D., Wabano, M., Russell, K., Enosse, L., & Young, N. L. (2014). Promoting resilience and wellbeing through an outdoor intervention designed for Aboriginal adolescents. *Rural and Remote Health*, 14(2523), 1–19.

Roth, J., Brooks-Gunn, J., Murray, L., & Foster, W. (1998). Promoting healthy adolescents: Synthesis of youth development program evaluations. *Journal of Research on Adolescence*, 8(4), 423–459.

Sallis, J. F., Cervero, R. B., Ascher, W., Henderson, K. A., Kraft, M. K., & Kerr, J. (2006). An ecological approach to creating active living communities. *Annual Review of Public Health*, 27, 297–322.

Sarkar, A., Hanrahan, M., & Hudson, A. (2015). Water insecurity in Canadian Indigenous communities: Some inconvenient truths. *Rural and Remote Health*, 15 (3354), 1–14.

Sayal, R., Bidisha, S. H., Lynes, J., Riemer, M., Jasani, J., Monteiro, E., ... Eady, A. (2016). Fostering systems thinking for youth leading environmental change: A multinational exploration. *Ecopsychology*, 8(3), 188–201.

Slade, M., Lowery, C., & Bland, K. (2013). Evaluating the impact of forest Schools: A collaboration between a university and a primary school. *Support for Learning*, 28(2), 66–72.

Thompson Coon, J., Boddy, K., Stein, K., Whear, R., Barton, J., & Depledge, M. H. (2011). Does participating in physical activity in outdoor natural environments have a greater effect on physical and mental wellbeing than physical activity indoors? A systematic review. *Environmental Science & Technology*, 45(5), 1761–1772.

Thurber, C. A., Scanlin, M. M., Scheuler, L., & Henderson, K. A. (2007). Youth development outcomes of the camp experience: Evidence for multidimensional growth. *Journal of Youth and Adolescence*, 36(3), 241–254.

United Nations. (2015). *Transforming our world: The 2030 Agenda for Sustainable Development*. New York: United Nations. Retrieved on April 11, 2018 from: https://sustainabledevelopment.un.org/post2015/transformingourworld/publication

United Nations Office on Sport for Development and Peace. (2017). *Sport and the sustainable development goals: An overview outlining the contribution of sport to the SDGs*. New York: United Nations Office on Sport for Development and Peace. Retrieved on January 23, 2018 from: www.un.org/sport/sites/www.un.org.sport/files/ckfiles/files/Sport_for_SDGs_finalversion9.pdf

Waite, S., Goodenough, A., Norris, V., & Puttick, N. (2016). From little acorns ...: Environmental action as a source of well-being for schoolchildren. *Pastoral Care in Education*, 34(1), 43–61.

Williams, D. R., & Dixon, P. S. (2013). Impact of garden-based learning on academic outcomes in schools: Synthesis of research between 1990 and 2010. *Review of Educational Research*, 83(2), 211–235.

Willox, A. C., Harper, S. L., Ford, J. D., Edge, V. L., Landman, K., Houle, K., ... Wolfrey, C. (2013). Climate change and mental health: An exploratory case study from Rigolet, Nunatsiavut, Canada. *Climatic Change*, 121(2), 255–270.

Young, N. L., Wabano, M. J., Burke, T. A., Ritchie, S. D., Mishibinijima, D., & Corbiere, R. G. (2013). A process for creating the Aboriginal children's health and well-being measure (ACHWM). *Canadian Journal of Public Health*, 104(2), e136–e141.

Chapter 11

Indigenous peoples, sport and sustainability

Dan Henhawk and Richard Norman

Introduction

Sport in the Indigenous context of Canada is a complex and diverse cultural space. It is a space rooted in modernist, Eurocentric philosophies and values, and deeply connected to the history of settler colonialism that sought to erase Indigenous physical cultural practices and the philosophies that informed them. Within this space, Indigenous peoples' participation has been marked by struggles to gain access to a colonized terrain that worked to normalize racialized notions of Indigenous inferiority and European superiority. This history, in turn, informs both the ostensible need for "development" in Indigenous communities as well as the ways in which sport has been integrated into such development programs. Indeed, the emergence of contemporary Sport for Development (SFD) programming directed at Indigenous communities – including, for example Right to Play's Promoting Life skills in Aboriginal Youth Program (PLAY) – are in many ways part of this history, and bespeak broader questions about power, resistance, and the reproduction of cultural values and societal norms that muddy the already complex landscape of Indigenous identities and cultural resurgence.

Within this milieu of sporting practices and socio-cultural tensions exist instances of Indigenous communities mobilizing to revitalize traditional land-based physical cultural practices built upon an Indigenous understanding of *relationship* (Wilson, 2008). Such a notion of relationship articulates an Indigenous philosophy of how to exist in the world as human beings, in a manner that often differs from Eurocentric conceptions of development and that may better account for "sustainable" relationships with the land and "traditional" culture. These tensions between sport, development and sustainability in turn highlight broader points of conflict surrounding Indigenous philosophies of relationship and Western notions of sustainable development.

In this chapter, we aim to address some of these tensions and challenges regarding Sport for Development and environmental sustainability in the Indigenous context. It is our contention that SFD programs in the Indigenous context are often problematic given the role of sport within histories of colonization and the potential of sport to perpetuate neo-colonialism. We further argue that SFD is often situated within notions of modernity in ways that make it difficult for such approaches to appreciate sustainability in the Indigenous context. We also suggest that, at this juncture, the discussions surrounding Indigenous knowledge within such programs create more questions than answers because of the fragmentation of knowledge that has occurred through colonization. As such, we attempt to privilege Indigenous perspectives that expose the tensions within Western knowledge and practices, and further Grande's (2011) call for a "resurrection of intellectualism and a resuscitation of the dialectic" (p. 42) that opens space for dialogue and criticism of Western values and knowledge that have become normalized. As such, this chapter attempts to highlight some of the questions, challenges and tensions that exist between Indigenous knowledge, sport and notions of environmental sustainability.

Before proceeding further, it is necessary to offer a note on terminology. A major challenge in discussing Indigenous peoples and knowledge is how to appropriately identify and represent a diverse group of people whose lives and cultures have been affected by colonialism. In this text, we use the term "Indigenous peoples" in reference to those who self-identify as Indigenous and to those cultures that existed on the North American continent prior to the arrival of European settlers. As Smith (2012) articulates, the term "Indigenous peoples" has allowed for a collective voice for those who:

> share experiences as peoples who have been subjected to the colonization of their lands and cultures, and the denial of their sovereignty by a colonizing society that has come to dominate and determine the shape and quality of their lives.
>
> (p. 39)

It is crucial that the term not be construed to mean that Indigenous cultures are monolithic or that contemporary Indigenous experiences are akin to the popular perception of an impoverished group of peoples that reside solely on reserves in the remote northern geographic regions of Canada. Instead, our usage is intended to situate Indigenous peoples within a colonial reality that has critical implications for discussions of sport and environmental sustainability. As such, it must be stated that any criticism within this text comes from an Indigenous scholarly standpoint that starts from a place of opposition and critique of colonialism inasmuch as it draws from specific cultural knowledge.

Another challenge within the Indigenous context is related to how to discuss and contextualize sport. As Kidd (2008) discusses, international development through sport has a long history, and most recently has been conceptualized as a "movement" known as Sport for Development and Peace

(SDP). In an overview of the history and landscape of this movement, Kidd makes a distinction between sport development (SD) and Sport for Development. He argues that SD is mostly connected to sport organizations that take as their goal to develop the capacity of sports in various contexts, whereas SFD is primarily concerned with utilizing sport as a vehicle to advance broad social development goals. He states, however, that the defining line between SD and SFD is "often blurred in rhetoric and practice" (p. 373). This point is especially pertinent within the Indigenous context, where SD and SFD have often occurred simultaneously, and with similar goals of social development. Further, while the provision and expansion of sport opportunities, along with other forms of Eurocentric leisure, was conducted ostensibly as a way to support the "development" of Indigenous people, sport was also being utilized to supplant Indigenous cultural practices (Forsyth & Wamsley, 2006; Fox, 2006, 2007) with the implicit and explicit goals of assimilation. It is thus important to acknowledge that Indigenous critiques do not necessarily make a hard distinction between SD versus SFD because both have been – and can be – complicit in colonial practices. With these caveats in mind, the chapter proceeds by offering a brief word on the historical use of SFD in the Indigenous context before tracing these histories through the contemporary context. We then offer a discussion of sport and modernity and examine how modernist underpinnings continue to inform SFD and sustainable development in Indigenous communities. The chapter concludes by providing some insights into how sustainable relationships through sport – informed by Indigeneity – might take form in future.

A brief history of sport and SFD in the Indigenous context

To understand Indigenous critiques of SFD, it is crucial to have at least a basic understanding of the history of European conceptualizations of leisure and physical cultural practices and the ways in which these were implicated in the project of colonization. Leisure is a European concept that is broadly discussed in relation to concepts of work, time and freedom (Heintzman, 2007). Within the numerous histories of colonization, Indigenous peoples, their physical cultural practices and the meanings associated with such activities were often misinterpreted through a European lens that privileged Eurocentric notions of leisure, as well as work, the use of time, and philosophies related to progress and modernity (Fox, 2007). For example, Fox (2007) notes that the definition of "game" was often left as self-evident; Indigenous practices were studied through a Eurocentric lens and then classified as "games of chance and skill or dexterity" (p. 219) while misinterpreting or dismissing any ceremonial and spiritual significance attached to such practices.

Such misinterpretation of Indigenous cultures was often accompanied by moral and racist judgments about Indigenous cultures that led "to repression of Indigenous practices and governance structures, to forced-labour in the service of Eurocentric leisure practices, and to commodifica-tion and objectification of Indigenous peoples and practices" (Fox, 2007, p. 219). In Canada, such judgments resulted in the creation of the Indian Act in 1876 and the Department of Indian Affairs in 1880 which, in turn, led to the banning of practices that did not fit within the government's goals of expanding capitalism (Paraschak, 1998). For example, the Potlatch, a gift-giving ceremony that was part of the governance practices for the Indigenous peoples on the west coast of Canada, was banned for this reason. At the same time, while cultural practices were banned, European leisure and sport practices were promoted. It was also during this era that sports and other forms of European leisure were introduced to Indigenous children within the Residential School system, which was itself created for the purpose of assimilating Indigenous children into Canadian society. Residential Schools have since been revealed as the site of egregious abuses and as an institutionalized form of cultural genocide. These histor-ies illustrate how colonial mindsets, policies and practices violently dis-missed Indigenous physical cultures, and in turn raise questions about the notion of Canada as a bastion of democracy when it in fact utilized Euro-centric leisure practices as a weapon of colonization.

For Indigenous peoples who voluntarily participated in the burgeoning landscape of Euro-Canadian sport, Paraschak (1989a, 1989b) notes that such experiences were most often marked by encounters of racism, exploitation and ethnocentric distortion. For example, during the Canad-ian Lacrosse Tours of 1876 to 1883, Indigenous peoples were exploited and put on display for their skills and to the benefit of those who exploited them. Such Indigenous experiences illuminate the contested nature of sporting practices whereby Indigenous peoples "acquiesced to, resisted, or accommodated the imposed expectations" (Paraschak, 1998, p. 122) of colonial Canadian society. As such, sport has been the site for important questions regarding power relations between Indigenous people and the Canadian Government, about the hegemony of positioning Euro-centric leisure as natural and legitimate, and about the agency of individ-uals and communities who attempted to engage in sport under their own terms.

As an example, in 1970, the Government of Canada, through the Depart-ment of National Health and Welfare, released "A Proposed Sports Policy for Canadians" (Paraschak, 1995) and in 1972, this policy led to the develop-ment of the Native Sport and Recreation Program (NSRP). This, in turn, was administered by Fitness and Amateur Sport, the unit responsible for sport, physical fitness and recreation at the federal level. The NSRP was a funding program to support various sport and recreation projects in Indigenous com-munities and was specifically "aimed at raising performance levels to the point where Native athletes could compete alongside other Canadians in elite competitions, while also providing services to a disadvantaged population" (Paraschak, 1995, p. 1). However, as Paraschak (1995) notes, government officials maintained ethnocentric assumptions about the "'legitimate' nature of sport, the rationale for providing government-funded sport opportunities and the relationship between sport and Native politics" (p. 4). Government officials operated under the assumptions that Euro-Canadian sports were the only legitimate form of physical activity, that competitive sports were the desired goal of physical activity participation and that Indigenous peoples aspired to participate in the Euro-Canadian sport system. Not surprisingly, such assumptions were challenged by various Indigenous leaders who utilized their funds from the NSRP to create "all-Native" sport competitions. These competitions reflected a viewpoint within the Indigenous community that such competitions could stand as legitimate alternatives to the mainstream sport system and were even necessary for sport development (though not necessarily Sport for Development) in the Indigenous context. As a result, these competitions, and an all-Native sport system, emerged as a response to government actions to control Indigenous participation in sport, and created a space whereby Indigenous peoples could access sporting opportunities, define the eligibility requirements to participate, and control the values and cultural elements to be privileged and incorporated into the competitions.

The creation of the all-Native sport competitions had some positive outcomes; Indigenous peoples gained access to sport on their own terms where otherwise they might not have had the opportunity. For example, women in the Six Nations of the Grand River community in Ontario, were able to participate in "All-Indian" sport competitions that created a safe space and resulted in empowering experiences (Paraschak, 1990). There is also the example of the North American Indigenous Games, a multi-sport competition in which Indigenous peoples have infused elements of Indigenous culture and values.

At the same time, and complicating the notion of sport as an inherent means of self-determination, sport also became a way for Indigenous people to survive amidst the violence of colonialism, particularly within Residential Schools. Residential Schools aimed to indoctrinate Indigenous children with the values and belief systems of Euro-Christian Canadian society; during this time, many Indigenous children were subjected to multiple forms of physical and emotional abuse and cultural erasure. The schools were part of Canada's broader "Aboriginal policy" that had the central goals of eliminating Indigenous governance, terminating historical treaties and removing Indigenous peoples' rights. Ultimately, the assimilatory aims of such policies were to "cause Aboriginal peoples to cease to exist as distinct legal, social, cultural, religious, and racial entities in Canada" (Truth and Reconciliation Commission of Canada, 2015, p. 1). It should come as no surprise, then, that many Indigenous cultures and languages have become irrevocably fractured, leaving individuals and communities to deal with the effects of such prolonged efforts to destroy Indigenous cultures.

Through an exploration of Residential School survivors, Forsyth (2013) found that sport and games played a crucial role in their ability to cope with the devastation of being removed from families and subjected to abuse and assimilation. She also noted that while sports provided Indigenous peoples with "avenues for positive self-expression and identification … the techniques used to monitor and control their bodies also transformed Aboriginal physical practices" (p. 32). Such programs were also highly gendered and often only available to boys. Using notions of disciplinary power and the internalization of self-control, Forsyth (2013) posits that sport and games within the Residential School system "can be understood as forms of discipline because they provided a clear set of methods and principles to inculcate a new docility into the pupils – a docility that would presumably facilitate their integration into mainstream society" (p. 21). As such, systems of sport, the disciplining of the body and the erasure of Indigenous cultural practices within Residential Schools – including those related to sport and play – raise significant questions about systems of control, power and hegemony, and the effect on Indigenous peoples' agency and autonomy over their participation in sport and other forms of leisure activity.

Contemporary SFD programming in Indigenous communities

It is against this backdrop, we argue, that many contemporary conceptualizations and practices of Sport for Development should be viewed. In Canada, SFD activities occurring in the Indigenous context include initiatives by national multisport service organizations (MSOs), national and provincial sport organizations (NSOs, PSOs), provincial and territorial Aboriginal sport bodies (PTASBs), corporations, and in some cases by Indigenous communities themselves. It is a varied landscape where partnerships are often developed to implement SFD initiatives. For example, the Aboriginal Sport Circle, a federal MSO that provides political advocacy for sport and recreation, may work with other MSOs and PTASBs to develop capacity for the provision of sport. In another instance, Right to Play, a global leader in international SFD for nearly two decades, implements programs that utilize games, play-based activities and sport to promote health and educational social development within Indigenous communities. Specifically, Right to Play currently delivers a program called Promoting Life-skills in Aboriginal Youth (PLAY) that partners with Indigenous communities and various organizations to train youth mentors to deliver "play-based programs that promote healthy living, healthy relationships, education and employability life-skills" (Right to Play, n.d.). NSOs and PSOs also engage in SD activities by organizing various initiatives for their respective sports. These initiatives are made possible through the Sport Support Program, the funding program to support sport and athlete development in Canada. Any initiatives by these respective organizations are thus subject to the objectives of the current Canadian Sport Policy which was first announced in 2012 and is effective until 2022.

Within this context PTASBs, the organizations that represent the interests of Indigenous peoples at the provincial level, deliver programs that support SD and SFD objectives, or they work in partnership with other MSOs or PSOs. The PTASBs are also part of a collective that is represented by the Aboriginal Sport Circle, the national organization that represents Indigenous interests regarding sport and recreation development to the federal government. It is important to note that PTASBs also provide support for Indigenous communities to train for and attend national "all-Native" championships such as the National Aboriginal Hockey Championships, the Canadian Native Fastball Championship and the North American Indigenous Games.

Taken as a whole, these initiatives point to the significant current deployment of sport to support the "development" of Indigenous peoples in Canada. At the same time, and in recognizing the history of colonialism and genocide, such programs call for critical analysis. While a plethora of approaches can be taken, in the remainder of this chapter we discuss two issues worthy of such attention when critiquing SFD from an Indigenous perspective: modernity and environmental sustainability.

Sport and modernity

Sport resembles, replicates and mimics our broader social and cultural landscape (Besnier & Brownell, 2012; De Wachter, 2001), and as such modern sport is implicated in the project of modernity (Clevenger, 2017). Modernity here is defined as the privileging of notions of equality, democratic systems and capitalistic ideologies, in which merit and performance are key instruments driving a push towards supposed egalitarian arrangements. For its part, modern sport – in contrast to earlier forms of sport – is often characterized by "rationalization, standardization, secularization, specialization, quantification, and records" (Breivik, 1998, p. 107). There are clear expressions of the measurement of performance, winners and losers can be clearly marked, and excellence quantified with exacting precision. Sport is thus compatible with modernist thinking, as understood in the following:

> One's place in society is not inherited at birth: it must be earned. He or she who performs will be held in esteem. But performance means something only if it can be measured against that of others: everyone is thus in constant competition with everyone else.
>
> (De Wachter, 2001, p. 93)

As a result of such thinking, the presumed transition from folk/traditional sports towards those deemed to be modern has come (often unproblematically) to be seen as simply a matter of historical phases, in which traditional physical cultures (including those of Indigenous peoples) have been overwritten with and by modern sporting forms (Besnier & Brownell, 2012, p. 448). In the Canadian context, the sport and recreation practices of many Indigenous groups were viewed as primitive, owing to a lack of standardized rules or clear contest, and therefore requiring modernist improvement. As Michael Heine (2013) recounts, games and sport of the Inuit and Dene primarily reflected the kinds of land-based practices that tested technical skills, strength and endurance required for daily life. In this context, sport and games "served to extend cultural logic of practice that focused on cooperation in subsistence production and other domains of social life" rather than strictly on the "symbolic validation of the outcome of the games contest" (Heine, 2013, p. 164). In contrast, modern sport aimed to override such approaches in deference and service to hierarchical and competitive relations of power and, even more poignantly, to colonialism and globalization (Besnier & Brownell, 2012; Jonasson, 2014; Roche, 2000). Attempts were made to supplant and rearrange traditional physical cultures in order to ensure privilege, hierarchy and dominance.

In sport, the "leveling" required for establishing such rules and regulations – and creating a fair, equitable and equal method of evaluation – tied sport to the tenets of modernity in ways that were compatible with capitalism. Modernity's democratic principles also entrenched equality within concepts of "fairness" that were central to sport, through the language of a "level playing field," performance-based measurement, and ideologies and narratives of success versus failure, in which winners could be exalted and losers chastised for their inadequacies. Through mastery of uncontrollable conditions – weather, infrastructure, physiology, competitors – sport exemplified both control and chance in ways compatible with the "illusion of modernity" (De Wachter, 2001, p. 95).

These perspectives are crucial to understanding current practices of SFD. Particularly from an Indigenous perspective, contemporary SFD is significantly limited by its overarching inability and/or unwillingness to escape from or reject discourses of modernity and modernist development. While there have been important calls for Indigenous knowledge in SFD (see Mwaanga & Mwansa, 2013), notions of modernity still tend to underpin the conceptualization of development throughout the SFD field. This is particularly the case when it comes to capitalism and its effects on the environment. The "promise" of development for Indigenous peoples in and through sport and SFD largely remains the promise of modernity and modern life. It is one in which Indigenous people can access the boons of modern capitalism by exploiting, taming and overcoming the natural world and, in so doing, overcoming the limits of their traditional cultures. From this perspective, SFD programs still tend to pride themselves on facilitating, and even "allowing," Indigenous people to claim the benefits (material, spatial and discursive) of full participation in contemporary society, all the while keeping the logic and practice of capitalist exploitation of the environment firmly in place (see Millington et al., 2019). That such promises are both the result and continuation of colonialism and colonial practices represents phenomena and issues that are still too rarely discussed in the field of SFD, and therefore call for ongoing critical analysis.

Sport and sustainability

In the face of such modernist exploitation, the notion of environmental sustainability assumes renewed importance. However, the notion of sustainability is itself far from neutral, historically, politically or socially. Sustainability has also often been a project of modernity, one in which a focus on preserving a harmonious way-of-being aligns with the romanticism, or even pathologizing, of Indigeneity. In this way, while sustainability's search for a different way of doing and being with the physical world may align with Indigenous cultures that are noted for stronger ties to the environment, "the very notion of sustainability is embedded in an essentially modern framework, entailing a number of contradictions and paradoxes, which can be interpreted as epistemic and normative diversions and obstacles" (Benessia et al., 2012, p. 75). Indeed, many contemporary notions of sustainability recall modernist understandings of progress and value because they share a belief in humans controlling nature. Just as control over Indigenous peoples was central to the mission of settler colonialism, so too was dominion over the natural world. For some, the very idea of sustainable development is predicated on such human intervention into nature. As Rist (2002) argues:

> Development consists of a set of practices, sometimes appearing to conflict with one another, which require – for the reproduction of society – the general transformation and destruction of the natural environment and of social relations. Its aim is to increase the production of commodities (goods and services) geared, by way of exchange, to effective demand.
>
> (p. 13)

Further, from a postcolonial perspective, Gidwani (2008) notes that land deemed to be "unproductive," "uncultivated," "idle" – as judged through the lens of modernity – has long served as justification for colonial interventions that could make "rational" use of it. In this way, sustainable development through modernization has often entailed a need to enact control over both humans and the landscape they occupy (Gasteyer & Butler Flora, 2000). As Gasteyer and Butler Flora (2000) recount, there is a pattern to such modernizing practices in a colonial context, whereby settlers move into a new landscape and (violently) claim ownership, establishing dominion over "wild," "harsh" and "unsettled" landscapes:

> Marshes are tiled and drained to create farm land for growing straight rows of corn, gains, or legumes; deserts are irrigated to grow horticultural crops the year rand. Rangeland is confined by fences, rather than driven by nomadic herders ... Even for significant parts of the subjugated indigenous population, the domination of nature as defined by the colonizers could come to symbolize progress, modernity, and rationality.
>
> (p. 129)

Such remaking of the environment was central to both colonial and development projects in which land and peoples were deemed to be in need of civilization and increased productivity (Gasteyer & Butler Flora, 2000). Rhetorical commitment to sustainable development does not necessarily overcome such traditions of exploitation.

The notion of environmental sustainability therefore invites critical scrutiny in order to examine and even expose its modernist underpinnings, while also challenging any homogeneous, single meaning of development. The critical task then becomes how to consider a range of tendencies, values and behaviors concerned with environmental stewardship and resilience, while exploring "the potential to maintain the long-term well-being of communities based on social, economic, and environmental requirements of present and future generations" (Cutter, 2014, p. 73). Admittedly, this is no small chore. Western scientific approaches to sustainability still predominate, and tend to limit contributions from Indigenous sources of knowledge, epistemologies and practices of inquiry. Indeed, the very notion of sustainability can do an injustice to Indigeneity by "rendering Indigenous peoples as anachronistic sources of insights, information and knowledge that can be used by science to produce authoritative, authentic and useful universal knowledge in the present, for the future" (Johnson et al., 2016, p. 2).

Overall, this challenge of pursuing sustainability while critiquing its modernist residue represents something of a crossroads for SFD research and researchers. On the one hand, there is a growing body of literature within mainstream sport sociology which explores how settler colonialism and development-as-modernization overlap with global sport in ways that become the basis of much SFD activity (see Darnell, 2010; Darnell & Hayhurst, 2011; Hayhurst, Giles & Wright, 2016). Such criticisms have only become more significant given the increasingly pervasive discourses regarding sport's ability to contribute to meeting the goals of international development, particularly environmental sustainability. This is perhaps best illustrated in the United Nations (2016) Sustainable Development Goals that not only foreground environmentalism as a key development issue, but have attached sport to the achievement of all 17 goals, calling it "an important enabler of sustainable development" (Article 37).

What is arguably missing from this growing field of critical SFD researchers, though, is a central place for Indigenous voices, cultures and ways of thinking and being. The environmental issues tied up in SFD are of particular relevance – and concern – within an Indigenous context, particularly as SFD programs continue to be positioned as development strategies for Indigenous peoples in Canada. In this way, sustainability and/as Indigenous development is (again) being used to justify capitalist exploitation of natural resources. A particularly illustrative example is the ways in which companies in the oil, mining and gas industry fund SFD programming in northern Canada as a means of offsetting the deleterious social and environmental impact of extractivism, a trend that may align more closely with "greenwashing" and public relations strategies for corporations than a genuine commitment to social and environmental justice (see Millington et al., 2019). A fuller appreciation of these processes from an Indigenous perspective is still needed. While sport sociologists like Millington and Wilson (2013) have expressed recent concern over how the sporting industry is driven by a "modernizing project" that entrenches a division between humans and natures, and the belief that "human 'progress' can be equated with 'mastering nature'" (p. 143), Indigenous perspectives have been making such claims for generations.

In sum, in the face of critical assessments of sustainable development in and through sport, the question remains as to what extent – or whether at all – sustainable development through sport and SFD is possible, and how this could be better rooted in, and respectful of, Indigeneity and Indigenous peoples. In the final section of this chapter, we ruminate on what kind of relationships might be possible and necessary to support Indigenous-led sustainability in and through SFD.

Conclusion: Indigenous approaches to sustainable sport

While sport has recently been cast as an agent of sustainable development on an international scale by organizations like the United Nations, Indigenous leaders have long perceived sport as a means to bolster social development and to demonstrate their cultural distinctiveness, with physical cultural practices offering a means to "highlight the great diversity among North American Indigenous peoples, to create a sense of unity from this diversity and to promote the resurgence of Indigenous cultures and cultural identities in non-Indigenous society" (Forsyth & Wamsley, 2006, p. 304). This occurred despite the complicity of sport with colonialism and modernist development in ways that worked to undermine Indigenous culture and self-determination. In recognizing this, we see an opportunity for the current field of Sport for Development to acknowledge and support Indigenous self-determination, particularly in ways that are compatible with more sustainable environmental relations. Doing so will require attending to the tensions and contradictions of sport discussed in this chapter. One of the most important of these tensions is that the types of sport being promoted through SFD, supported by government and corporate sponsors, and even called for within the recommendations of the Truth and Reconciliation Commission, continue to take place on colonized land. The North American Indigenous Games are perhaps an example of this, in the sense that traditional cultural practices of Indigenous peoples have been interrupted, and rendered inaccessible through (neo)colonial practices of settler colonialism, practices that are now attempting to be reconciled through Euro-Canadian sport development. Conversations about land, as both a material and discursive basis for contemporary sport, need to be had, and in so doing, can help to open up critical conversations about Indigenous approaches to sport and sustainability.

Berkes (2018) argues that "Traditional Ecological Knowledge" (TEK) may offer a means to promote better understandings of ecological processes and relations with the environment through knowledges, practices and a "sense of place." TEK is viewed as a "cumulative body of knowledge, practice, and belief, evolving by adaptive processes and handed down through generations by cultural transmission, about the relationship of living, being with one another and with their environment" (Berkes, 2018, p. 7). Such an approach may go some distance in responding to Eurocentric notions of development-as-modernization by embedding environmental knowledge in local culture, promoting the importance of community, and attempting to break down the barriers between nature and culture so as to promote more sustainable relationships between people and nature.

To push forward such an agenda, there is also a critical need to emphasize how an Indigenous notion of *relationship* can inform the development of sustainable sport. Wilson (2008) puts forward the idea of a "relational way of being" (p. 80) that is predicated on the foundational belief that:

> knowledge is relational. Knowledge is shared with all of creation. It is not just interpersonal relationships ... but it is a relationship with all of creation. It is with the cosmos, it is with the animals, with the plants, with the earth that we share this knowledge.
>
> (Wilson, 2008, p. 74)

He argues that "rather than viewing ourselves as being *in* relationship with other people or things, we *are* the relationships that we hold and are part of" (Wilson, 2008, p. 80). Such an orientation has the potential to expand current notions of sustainable development, as well as challenge how we envision Sport for Development. Such a change would shift the priority of SFD towards attending to a "relationship way of being" in the world, one that might prove to be healing and more sustainable in the long term.

References

Benessia, A., Funtowicz, S., Bradshaw, G., Ferri, F., Ráez-Luna, E. F., & Medina, C. P. (2012). Hybridizing sustainability: Towards a new praxis for the present human predicament. *Sustainability Science*, 7(S1), 75–89. doi:10.1007/s11625-011-0150-4

Berkes, F. (2018). *Sacred ecology* (4th ed.). New York: Routledge.

Besnier, N., & Brownell, S. (2012). Sport, modernity, and the body. *Annual Review of Anthropology*, 41, 443–459. doi:10.1146/annurev-anthro-092611-145934

Breivik, G. (1998). Sport in high modernity: Sport as a carrier of social values. *Journal of the Philosophy of Sport*, 25(1), 103–118. doi:10.1080/00948705.1998.9714572

Clevenger, S. M. (2017). Sport history, modernity and the logic of coloniality: A case for decoloniality. *Rethinking History*, 21(4), 586–605. doi:10.1080/13642529.2017.1326696

Cutter, S. L. (2014). Building disaster resilience: Steps toward sustainability. *Challenges in Sustainability*, 1(2), 72–79. doi:10.12924/cis2013.01020072

Darnell, S. (2010). Power, politics and "sport for development and peace": Investigating the utility of sport for international development. *Sociology of Sport Journal*, 27, 54–75.

Darnell, S., & Hayhurst, L. (2011). Sport for decolonization: Exploring a new praxis of sport for development. *Progress in Development Studies*, 11(3), 183–196. doi:10.1177/146499341001100301

De Wachter, F. (2001). Sport as mirror on modernity. *Journal of Social Philosophy*, 32(1), 90–98. doi:10.1111/0047-2786.00081

Forsyth, J. (2013). Bodies of meaning: Sports and games at Canadian residential schools. In J. Forsyth & A. Giles (Eds.), *Aboriginal peoples and sport in Canada: Historical foundations and contemporary issues* (pp. 15–34). Vancouver, BC: UBC Press.

Forsyth, J., & Wamsley, K. B. (2006). "Native to native ... we'll recapture our spirits": The world indigenous nations games and North American Indigenous Games as cultural resistance. *International Journal of the History of Sport*, 23(2), 294–314. doi:10.1080/09523360500478315

Fox, K. M. (2006). Leisure and Indigenous peoples. *Leisure Studies*, 25(4), 403–409. doi:10.1080/02614360600896502

Fox, K. M. (2007). Aboriginal peoples in North American and Euro-North American leisure. *Leisure/Loisir*, 31(1), 217.

Gasteyer, S. P., & Butler Flora, C. (2000). Modernizing the savage: Colonization and perceptions of landscape and lifescape. *Sociologia Ruralis*, 40(1), 128–149.

Gidwani, V. (2008). *Capital, interrupted: Agrarian development and the politics of work in India*. Minneapolis, MN: University of Minnesota Press.

Grande, S. (2011). Confessions of a full-time Indian. *Journal of Curriculum and Pedagogy*, 8(1), 40–43.

Hayhurst, L., Giles, A., & Wright, J. (2016). Biopedagogies and Indigenous knowledge: Examining sport for development and peace for urban Indigenous young women in Canada and Australia. *Sport, Education and Society*, 21(4), 549–569. doi:10.1080/13573322.2015.1110132

Heine, M. (2013). Performance indicators: Aboriginal games at the Arctic Winter Games. In J. Forsyth & A. Giles (Eds.), *Aboriginal peoples and sport in Canada: Historical foundations and contemporary issues* (pp. 160–181). Vancouver, Canada: UBC Press.

Heintzman, P. (2007). Defining leisure. In R. McCarville & K. MacKay (Eds.), *Leisure for Canadians* (pp. 3–12). State College, PA: Venture Publishing.

Johnson, J. T., Howitt, R., Cajete, G., Berkes, F., Louis, R. P., & Kliskey, A. (2016). Weaving Indigenous and sustainability sciences to diversify our methods. *Sustainability Science*, 11(1), 1–11. doi:10.1007/s11625-015-0349-x

Jonasson, K. (2014). Modern sport between purity and hybridity. *Sport in Society*, 17 (10), 1306–1316. doi:10.1080/17430437.2014.850807

Kidd, B. (2008). A new social movement: Sport for development and peace. *Sport in Society*, 11(4), 370–380.

Millington, B., & Wilson, B. (2013). Super intentions: Golf course management and the evolution of environmental responsibility. *Sociological Quarterly*, 54, 450–475. doi:10.1111/tsq.12033

Millington, R., Giles, A., Hayhurst, L., van Luijk, N., & McSweeney, M. (2019). "Calling out" corporate redwashing: The extractives industry, corporate social responsibility and sport for development in Indigenous communities in Canada. *Sport in Society*. doi:10.1080/17430437.2019.1567494

Mwaanga, O., & Mwansa, K. (2013). Indigenous discourses in sport for development and peace: A case study of the Ubuntu cultural philosophy in Edusport Foundation, Zambia. In N. Schulenkorf & D. Adair (Eds.), *Global sport-for-development* (pp. 115–133). London: Palgrave Macmillan.

Paraschak, V. (1989a). Native sport history: Pitfalls and promise. *Canadian Journal of History of Sport*, 20(1), 57–68.

Paraschak, V. (1989b). The Native Sport and Recreation Program, 1972–1981: Patterns of resistance, patterns of reproduction. *Canadian Journal of History of Sport*, 26(2), 1–18.

Paraschak, V. (1990). Organized sport for native females on the Six Nations Reserve, Ontario from 1968 to 1980: A comparison of dominant and emergent sport systems. *Canadian Journal of History of Sport*, 21(2), 70–80.

Paraschak, V. (1995). Invisible but not absent: Aboriginal women in sport and recreation. *Canadian Woman Studies*, 15(4), 71–72.

Paraschak, V. (1998). "Reasonable amusements": Connecting the strands of physical culture in Native lives. *Sport History Review*, 29, 121–131.

Right to Play. (n.d.). PLAY symposium. Retrieved from www.righttoplay.ca/en-ca/national-offices/national-office-canada/get-involved/play/

Rist, G. (2002). *The history of development: From Western origins to global faith.* New York: Zed Books.

Roche, M. (2000). *Mega-events and modernity: Olympics and expos in the growth of global culture*. London: Routledge.

Smith, L. T. (2012). *Decolonizing methodologies: Research and indigenous peoples* (2nd ed.). New York: Zed Books.

Truth and Reconciliation Commission of Canada. (2015). *Honouring the truth, reconciling for the future: Summary of the final report of the Truth and Reconciliation Commission of Canada* Ottowa, Canada: Truth and Reconciliation Commission of Canada. https://doi.org/9780660019857, 066001985X

United Nations. (2016). *Sport and the Sustainable Development Goals: An overview outlining the contribution of sport to the SDGs*. New York: United Nations.

Wilson, S. (2008). *Research is ceremony: Indigenous research methods*. Black Point, Canada: Fernwood Publishing.

Chapter 12

Co-transforming through shared understandings of land-based practices in Sport for Development and Peace

Victoria Paraschak and Michael Heine

Introduction

In 2015, the Truth and Reconciliation Commission of Canada (TRC) released its *Final Report*. The three-member Commission had spent six years travelling across Canada to document the stories of physical, sexual and cultural abuse experienced by Indigenous children forced into residential schools for over 150 years, with the last school closing as recently as 1996. The *Final Report* issued 94 calls to action (CTAs). Many Canadians are currently committed to reconciliation as promoted through the TRC Final Report. For example, the National Centre for Truth and Reconciliation (NCTR) that arose out of the activities of the TRC conducts public 'Imagine a Canada' campaigns to raise awareness of the residential school history, particularly within the Canadian educational system (NCTR, 2018a; see also Aitken & Radford, 2018; Smith & Taunton, 2018). The NCTR also established the 'Memorial Register' intended to honor and commemorate children who did not survive the residential school system and never returned home; completion of the register is dependent on input and suggestions from the Canadian public (NCTR, 2018b). The NCTR regularly issues open calls to interested community organizations across Canada to participate in these and other campaigns. Further, the Canadian Broadcasting Corporation (CBC) has set up the 'Beyond 94' monitoring program to report on the progress made in the implementation of the TRC's CTAs. As of June 2018, Beyond 94 reported that initiatives on 50 CTAs were in progress, initiatives on 10 CTAs were complete, but also that initiatives on 34 CTAs had not yet started (CBC, 2018). Notably, the two relevant Canadian federal government departments – Indigenous Services Canada and Crown-Indigenous Relations and Northern Affairs Canada (CIRNAC) – report that concerning the five CTAs relating to sport and Indigenous physical activity practices (numbers 87–91), budget allocations to enable the development of long-term Indigenous sport development programs will amount to Can$20 million over five years (CIRNAC, 2018a, 2018b).

Since the five CTAs pertaining to sports have thus proven influential in recent policy making as well as research development, we use them here as the starting point for our discussion regarding the utility of a Sport for Development and Peace (SDP) paradigm as applied to Indigenous physical activity practices in Canada, including, but also reaching beyond, contemporary sports. SDP initiatives for Indigenous communities in Canada have gained increasing attention (Hayhurst & Giles, 2013; Gardam, Giles & Hayhurst, 2017). This process preceded the publication of the TRC's findings, but it has accelerated since then. Studies investigating this development identify several foci of SDP initiatives targeted on Indigenous communities in Canada, in particular an emphasis on cross-cultural mentorship as well as on the importance of community engagement for successful program development (Gardam, Giles & Hayhurst, 2017, 34–35). On the other hand, the intent of such programs is frequently influenced by development objectives defined at political levels extending beyond the space of sport development and policy (Gardam, Giles & Hayhurst, 2017, 36). This is also true of projects that extend the scope of SDP projects beyond sports into the area of Indigenous cultural physical activity practices (PAPs) more generally, as we note below.

One of the key statements in the CTAs addressing the politics of Indigenous sports and PAPs in Canada, is contained in CTA 90. Section i of CTA 90 reads:

> 90) We call upon the federal government to ensure that national sports policies, programs, and initiatives are inclusive of Aboriginal peoples, including, but not limited to, establishing:
>
> i. In collaboration with provincial and territorial governments, stable funding for, and access to, community sports programs that reflect the diverse cultures and traditional sporting activities of Aboriginal peoples.
>
> (TRC, 2015, 336)

In reflecting on the unproblematic linkage between Indigenous[1] traditions and sport constructed in CTA 90.i, we argue here for a conceptualization of and support for community-focused physical activity practices that are built around Indigenous land-based ways of knowing and that extend beyond the space of sports. We adopt this perspective since CTA 90's positive emphasis on 'traditions' in the physical activities space allows us to establish a primary concern not with sports – their great positive impacts in Indigenous communities notwithstanding – but with those Indigenous physical activities whose cultural significance emerges from the ways in which they connect Indigenous peoples to the land.

Given the unquestionable dominance of contemporary sports as a cultural practice, the possibility for the emplacement of such understandings has to be recovered in at least partial distinction from sports' dominant cultural logic, by which we mean sports' extensive emphasis on the importance of outcome, inter-personal competition and winning. To map out such a recovery, we draw on a 'Strengths and Hope' perspective (Paraschak, 2013; Paraschak & Thompson, 2014) as an augmentation of the SDP paradigm. Leaving aside the potential of sports to contribute to 'peace,' notions of 'development' brought to bear on Indigenous cultures in Canada have, under conditions of colonialism, contributed to a highly problematic history. Our interpretation aligns with Lavalée and Lévesque's argument that colonization, in part, restricted Aboriginal peoples from 'engaging in their traditional physical practices and the lifelong lessons that stemmed from them' (Lavalée & Lévesque, 2013, 210). Citing the medicine wheel teachings as an example, they instead argue for legitimizing systems of meaning that emphasize balance and a 'wholistic view towards health [involving an] interconnectedness within and with all creation' (p. 214). In such a wholistic approach towards life and physical activity, sport is constituted as only one possible manifestation of physical activity practices. At the very least, in view of their great dominance on a global scale, sports reproduce a homogeneity (Donnelly, 1996; Wamsley, 2002) that does not necessarily articulate to Indigenous cultural – 'traditional' – expression at locally and regionally specific levels. We thus refocus SDP by emphasizing a complementary Strengths and Hope rather than a development perspective, and we argue that such strengths should be sought for and located in the inherency of Indigenous cultural identities emerging from physical activity practices connecting Indigenous cultures to the land in locally and regionally specific ways. Rather than being positioned as the object of sport's beneficial developmental intervention, Indigenous cultures can in this perspective insist on cultural strengths that inhere in their own land-based physical activity practices.

The point of this chapter is to investigate what advantages might accrue from this shift of perspective. We do this by first developing a framework for analysis informed by a Strengths and Hope perspective, with a particular focus on its potential for co-transformation. We then offer a brief survey of examples of potential linkages of physical activity practices to the land by examining land-based practices of the Dene and Inuvialuit, a general reference for some of the Indigenous cultures whose traditional land use areas extend across the three Canadian Territories, Nunavut, Yukon and the Northwest Territories (NWT). In suggesting aspects of a wholistic – 'inherent' – perspective, we seek to guide the discussion to a point beyond the exclusive focus on physical activity practices. Next we present some current examples of Indigenous land-based physical activity programs, investigating their SDP potential from a Strengths and Hope perspective, as well as a summary of research exploring benefits and concerns tied to time in the out of doors for Canadian youth in general. We then offer some comments, from a co-transformation point of view, on the potential contribution of Indigenous land-based PAPs to social development for Indigenous as well as non-Indigenous participants.

We end with some reflections on the political dimension of the land-based PAP model examined here for SDP initiatives developed in Canada, in the context of the ongoing displacement of Indigenous peoples.

Framework for analysis: Strengths and Hope perspective

The Strengths perspective, which originated in social work (Saleebey, 2013), emerged as an alternative approach to the Deficit perspective dominant in North American society. The Deficit perspective begins by identifying current problems and then draws on the skills of experts to address those challenges (Paraschak, 2013; Saleebey, 2013). SDP programs, for example, which provide 'experts' to 'develop' needed skills – usually mainstream sport-focused and/or leadership-based – in community members (Hayhurst & Giles, 2013, 509) align logically with a Deficit perspective. Conversely, the Strengths perspective assumes as a first principle that 'every individual, group, family, and community has strengths' (Saleebey, 2013, 17), and that thus the analysis always starts by identifying such strengths, tied to the issue under examination. For the northern Dene and Inuvialuit communities, whose example is discussed below, a Strengths perspective analysis would thus begin by identifying a key strength in those communities – the particular culturally based ways that Indigenous groups and communities view their relationship to the land (Heine, Andre, Kritsch & Cardinal, 2007).

A second principle of the Strengths perspective is that 'every environment is full of resources' (Saleebey, 2013, 20) which can be drawn upon to further existing strengths. We would add to this principle that the resources can, at times, also be used to generate new useful strengths. In Dene communities, for example, elders serve as key resources who could explain cultural understandings tied to the land and their links to PAPs. They could thus serve as resources to help Indigenous and non-Indigenous youth imagine possible connections they might develop towards the land, along with approaches they could take to link such understandings to meaningful PAPs.

The third principle of a Strengths perspective is that rather than operating as 'experts' or functioning as 'recipients,' all individuals work together (Saleebey, 2013) with an openness to co-transformation (Paraschak, 2013). Practices of hope (Paraschak, 2013) guide this process, which align with Jacobs (2005) depiction of 'hope' as being fundamentally grounded in relationships, leading to 'hope in' a preferred, shared vision, rather than an individual understanding of 'hope for' an individually specific outcome. Existing SDP programs, in part, potentially align with this conceptualization of hope. For example, the United Nations General Assembly Resolution 58/5, titled *Sport as a Means to Promote Education, Health, Development and Peace* (UN, 2003), lays out shared goals to which all SDP programs might aspire – including a focus on cultural bridging, dialogue, local needs and the entrenching of collective values. It states in part that it:

1.   Invites Governments, the United Nations, its funds and programmes, the specialized agencies, where appropriate, and sport-related institutions:

   (a)   To promote the role of sport and physical education for all when furthering their development programmes and policies, to advance health awareness, the spirit of achievement and *cultural bridging* and to *entrench collective values*;
      [...]
   (c)   To work collectively so that sport and physical education can present opportunities for solidarity and cooperation in order to promote a culture of peace and social and gender equality and to *advocate dialogue* and harmony;
      [...]
   (e)   To further promote sport and physical education, *on the basis of locally assessed needs*, as a tool for health, education, social and cultural development.

(UN, 2003, italics added)

Accordingly, UN Resolution 58/5 serves as a *resource for creating effective SDP programs* grounded in a Strengths and Hope perspective. It suggests that a spirit of cultural bridging and dialogue be brought to any process, as together all participants work towards collective goals addressing local needs.

This process becomes clearer when aligned with the Practices of Hope (Paraschak, 2013), a concept that draws on Jacobs' (2008) exploration of hospitality tied to the concept of hope. Jacobs (2008) spoke about a Benedictine-based concept of 'hospitality' to outline the process he connected to his understanding of 'hope.' In this approach, all individuals need to value one another, and to listen and share perspectives with each other with a commitment to understanding rather than persuasion, open to the possibility of co-transformation. These practices could ensure that reflexive attention is given to 'hope in' a shared community vision wherein all individuals are transformed. They might prompt Aboriginal and non-Aboriginal individuals to collectively work to (re-)envision their aspirations tied to physical activities in their lives. They could then reflexively work to transform current conditions in this desired direction. They could also identify, draw upon and enhance their collective strengths by using accessible resources flowing from existing unequal but complementary power relations (Paraschak, 2013, 232).

This becomes one potential approach administrators could adopt as they structure SDP programs with an intention of being co-transformed, of learning about the strengths of the community which they are entering as well as sharing the strengths they bring, while also generating and working towards a shared preferred future. This process aligns with the fostering of 'hope' described by Denise Larsen as 'the ability to envision a future in which we wish to participate' (Enright, 2014). This approach also aligns with recommendations about SDP programs (Nicholls, Giles & Sethna, 2010; Hayhurst & Giles, 2013) meant to address the criticism that such programs do not focus on the co-construction of knowledge about the program and its outcomes because they ignore the legitimacy of 'subjugated knowledges' (Nicholls, Giles & Sethna, 2010, 251).

A Strengths and Hope perspective could also be valuable in SDP programming given concerns about nature deficit disorder currently being expressed across North America, which is the recognition that mainstream youth are no longer spending time in the out of doors and thus are missing out on the many benefits that emerge when that relationship is fostered (Louv, 2008). Louv argues for the need to 'consider the strong cultural links to nature that already exist and can be built on' (2011, 138) as they are central to an understanding of our integral role within nature. Exposure to Indigenous understandings of a fundamental link to the land on which they live could greatly aid mainstream youth in such land-based cultural understandings. Louv raises the question: 'What happens to a [human] species that loses touch with its habitat?' (2008, 18). He then points out that the increasing lack of connectedness among young North Americans to the natural world means that we will eventually have fewer adult stewards able to argue for the need to address pressing environmental issues such as climate change, sustainability and reduced environmental harms because they will not have solidified their connectedness to nature, as children. Once again, this is a reason why SDP programs between Indigenous and non-Indigenous youth could benefit all youth who deepen their appreciation of Indigenous cultural teachings tied to the land. Finally, Kelly (2018) notes that 25% of Americans spend their entire day indoors, referring to them as the 'indoor generation.' SDP programs, building on strengths informed by Indigenous ways of knowing the land, could look to Indigenous elders as resources to counter current North American discourses which portray individuals as increasingly disconnected from the land. These programs could facilitate having all participants commit to adopting PAPs that draw upon and enhance a connectedness to the land – a possibility we explore through this chapter.

Physical activity spaces, 'sport' and Indigenous land-based activities

Undoubtedly, contemporary sports have come to play a significant positive role in Indigenous communities in Canada, but their cultural dominance creates specific challenges for land-based Indigenous PAPs. Sport's spatial constitution, for one, requires the construction of a synthetic space (the 'playing field') whose indispensable uniformity abstracts it from the environment within which it is embedded, and which alters the natural landscape – often causing environmental degradation – through that creation. Most sports require this uniform spatial abstraction to express their invariant logic on a global scale (Bale, 1993, 1994, 100–119; Guttmann, 2004).[2] Sports and their spaces, in this sense abstracted from their environment, also enfold a specific cultural context and positioning, since they carry their own – by and large uniform – values, significant practices and assumptions about appropriate forms of action, intentionality, interpersonal behavior and so forth. These can often be at significant variance with the value orientation of Indigenous PAPs, including many Indigenous (traditional) games, the set of Indigenous PAPs most immediately complementary to contemporary sporting practices (see, e.g., Heine & Scott, 1994).

By contrast, Indigenous connections to the land reproduced and confirmed through PAPs constitute an essential part of various sets of interrelated connectivities, in the immediate land-related sense, but also, and perhaps more importantly, within social and cultural systems of practice and meaning. We will elaborate on this argument by briefly considering some examples of northern Canadian Indigenous land-based PAPs. Our choice of examples is guided by our own familiarity, but the argument has applicability beyond the Canadian north. The Dene (Athapaskan), a group of Indigenous cultures whose traditional land use areas are located in parts of interior Alaska, Yukon Territory, the NWT and the northern regions of several Canadian provinces, offer a case in point. The land-based way of life of Dene family groups was closely linked to the characteristics of the northern environment. The adaptation pressures of the challenging subarctic climate and the reliance on seasonally fluctuating food resources (in particular moose and caribou) determined a life style involving considerable travel and physical exertion during certain seasons of the year (e.g., Helm, 2000; Heine, Andre, Kritsch & Cardinal, 2007, 67–128). Closely linked to the structure of subsistence production, social organization was highly flexible, emphasizing mutual cooperation and sharing on the one hand and an ethos of individual autonomy, and self-reliance on the other (Honigman, 1968). The emphasis on individual autonomy corresponded to an absence of permanent hierarchical structures, that is, to a 'disinclination for superordinate-subordinate systems of relationship' (Honigman, 1975, 560; cf. Christian & Gardner, 1977, 67–70).

This space of interrelated environmental and cultural factors influenced the internal logic of many of the Indigenous games played by the Dene, to focus on those PAPs conceptually most closely related to the domain of sports. Reflecting the high degree of spatial mobility required in subsistence production, many traditional games focused on the display of physical skills and athletic competency that functionally related to skills required in land-based subsistence production, at least as far as men's participation was concerned. But, articulating to the more egalitarian nature of interpersonal relations, the importance of relative, individuated achievement and winning was de-emphasized. An inventory of the games of the Dene could list many activities whose emphasis was on process more than on outcome, on inclusion rather than the demarcation of winners from losers (Heine, 2012). The appreciation of athletic achievement and the competent display of physical skills did not connote a related and inevitable emphasis on interpersonal competition, that is to say, on the comparative aspects of physical prowess characteristic of the modern system of competitive sports. The tendency not to 'stand out' in this particular field of activities was congruent with similar 'disinclinations' evident in the general social space (Heine & Scott, 1994; Heine, 2012).

A complementary example can be seen in the traditional games of the Inuit, historically also land-related, but at the same time emphasizing the importance of individual competition much more strongly. However, at a fundamental level, the same connectivities linked Inuit to the land through their cultural PAPs (see, e.g., Bennett & Rowley, 2004, 339–434). The elaboration of these systems of interrelationships varies at local and regional levels, but the argument's generalizability insists on the constitutive inherency of Indigenous cultures' relationships to the land, based on, and reproduced through, inter alia, specific PAPs, including traditional games. It is important to note that these cultural orientations are in evidence at northern Indigenous games festivals to this day (e.g., Heine, 2013).

There is an additional important point worth making here. In a wholistic perspective, the connections to the land that emerge from land-based PAPs also serve to reconnect Indigenous peoples not only to the land, but also to 'the social relations, knowledges and languages that arise from the land' (Wildcat, McDonald, Irlbacher-Fox & Coulthard, 2014, 1). Such knowledges fundamentally involve what Styres (2017, 41–59) refers to as 'storying the land,' that is, the expression and recovery of cultural knowledge about the land (see, e.g., Heine, Andre, Kritsch & Cardinal, 2007, 7–58) that informs, much as it is guided by, relationships to the land created through PAPs. Narrative practices and PAPs are thus interconnected in the creation of wholistic cultural meanings (e.g., Cruikshank, 1977, 2005; Styres, 2017). This perspective on the production of cultural meanings and identities also locates positions of authority with Indigenous participants as the keeper of cultural land-related knowledges, through and about PAPs; the role and significance of Indigenous elders, in particular, would be foregrounded in this perspective. Cultural knowledges and identities emerging at the intersection of PAPs and 'storying' could also offer meaningful perspectives to those Indigenous participants for whom land-based experiences are at present a matter of discontinuous experience or, particularly in urban settings, often not easily available.

Further, this perspective might also hold the potential to inverse the relationships often implied by SDP approaches, to a Strengths perspective where Indigenous partners as holders of cultural knowledge hold positions of authority based on this knowledge. Such projects might, in turn, hold the potential for co-transformation, with non-Indigenous participants being in the position of learners.

Perspectives for such co-transformation could, for example, be indicated by the traditional knowledge authority of Dene or Inuit community members who are land-based subsistence practitioners; they could – and do – serve as human resources who can explain their cultural understandings tied to the land. This would include associated links to PAPs (Redvers, 2016; Dechinta, n.d.). These stories, along with the other resources mentioned, all serve as a means by which Indigenous and North American youth can be helped to develop or enhance existing, meaningful links they might make to the land around them, as well as approaches they could take to link these understandings to physical activities. Deepening their awareness of and connectedness to the natural world could then potentially foster an enhanced commitment to protecting that environment through supporting sustainable development and/or addressing the issue of climate change. Throughout this process, all individuals would work in an engaged and respectful manner with, rather than on behalf of, the group being assisted, and toward 'hope in' a shared preferred future, maintaining an openness to co-transformation through that process. 'Hope in' this preferred future lies in a shared desire for reconciliation, as outlined in the Truth and Reconciliation *Final Report*:

> Reconciliation requires that a new vision, based on a commitment to mutual respect, be developed. It also requires an understanding that the most harmful impacts of residential schools have been the loss of pride and self-respect of Aboriginal people, and the lack of respect that non-Aboriginal people have been raised to have for their Aboriginal neighbours. Reconciliation is not an Aboriginal problem; it is a Canadian one.
>
> (TRC, 2015, vi)

These harmful impacts, if focused upon, continue to reinforce the dominant Deficit perspective evident in mainstream society. Instead, we argue for the adoption of a Strengths and Hope perspective, as an approach that more effectively aligns with the 'preferred future' of reconciliation.

On-the-land projects and educational policy

A wide range of programs are under development in educational institutions and in Indigenous communities and organizations across Canada that are designed to recover and strengthen the connections between Indigenous communities, in particular youth, and the land. Some of these programs are developed with an emphasis on mental health resilience; others are designed to foster and strengthen relationships between members of communities that often are exposed to severe stressors such as family and inter-gender violence, or substance abuse; others train in skills required in land-based subsistence production, and general knowledge about the local environment. They are unified by an underlying assumption about the inherency of land-based relationships for the construction of Indigenous identities.

In the Canadian north, a well-known example of such on-the-land programming in Yukon Territory is provided by the Jackson Lake Wellness and Recreation Team. A joint initiative by the Government of Yukon, Kwanlin Dün First Nation, and Health Canada, it provides healing experiences and tools for people to address issues 'related to residential school, trauma, addictions, depression, loss and grieving. The program complements culture- and land-based activities with clinical approaches' (Health Canada, 2016, 5). Key program components include the production of culturally relevant prevention-focused materials and activities, participant and family outreach support, as well as culturally supportive counseling and guidance (p. 6). The program admits Indigenous as well as non-Indigenous participants, and in this way adheres to the tenet of co-construction advocated in this chapter.

Similarly, in the Northwest Territory, one of the key programs extending across several communities is Project Jewel, funded by the NWT Government and Health Canada and managed by the Inuvialuit Regional Corporation. The project delivers on-the-land programs that establish safe and culturally relevant environments to encourage participants' healing while drawing on traditional cultural strengths. Themes of the land-based camps have included residential school trauma, family violence amelioration and the importance of traditional values. The project primarily works with residents in the five communities within the Inuvialuit Settlement Region (Health Canada, 2016, 5). Lastly, the Dechinta Centre for Research and Learning, a northern-led initiative associated with the University of Alberta, offers land-based educational experiences for Indigenous and non-Indigenous participants; educational programs are guided by northern leaders and elders in collaboration with university knowledge workers (Dechinta, n.d.).

In addition, the NWT, Nunavut and Yukon have all developed curriculum documents and instructional contents that make emphatic connections between various aspects of Indigenous culture and the land base. In the NWT, the *Dene Kede* curriculum document was developed through a two-year collaboration between elders, linguists, educators and NWT government representatives. A generic curriculum document for grades

K-6 was published in 1993 (NWT, 1993) and later expanded to include K-9. It serves as the structure used by community schools and cultural organizations to develop locally and culturally relevant educational materials. Focusing in particular on language acquisition, learner expectations defined in *Dene Kede* are divided into four quadrants, with one quadrant emphasizing relationships to the land in their connectedness to other aspects of Dene culture: 'the people – the spiritual world – the self' (NWT, 1993, xxxi). The initial document has since been used for the development of culturally relevant programs in many educational settings in the Dene communities.

The development of a curriculum document for Inuit culture, *Inuqatigiit*, was also initiated at the same time, and applied analogous principles of development (NWT, 1996) while focusing the document contents specifically on Inuit culture. Connections to the land are expressed in one of the three concentric foundational cycles that represent Inuit culture in this document; the 'cycle of the seasons' (NWT, 1996, 30–32) is related to the cycles of life, and of belonging (p. 38), thus offering a wholistic view of Inuit culture. After the separation of Nunavut from the NWT in 1999, the *Inuuqatigiit* curriculum was also approved for use in the Nunavut education system. Through the development of the Inuit Traditional Knowledge Project (Inuit Qaujimajatuqangit), educational and cultural organizations in Nunavut have since implemented a major Indigenous knowledge project that also focuses significantly on connections between culture, language and the land (Nunavut, 2007; McGregor, 2012). The framework is noteworthy for the fact that it provides the whole culture-based foundations for the educational system of an entire Territory (Nunavut, 2007, 5).

Addressing nature-deficit disorder

Focusing on traditional Indigenous land-based understandings and accompanying physical activity practices through SDP programs also holds promise for addressing concerns in North America that individuals are increasingly disconnected from nature and outdoor physical activities. As discussed, Louv (2005) was one of the first to call attention to, and ignite discussions around, youth and time spent (or not) in the outdoors. Coining the term 'nature-deficit disorder,' he discussed the increasing alienation of North American youth from time in the outdoors, prompted by socially constructed parental fears such as 'stranger danger' and perceived risks tied to outdoor activities, as well as an increasing preference by youth for 'spending time indoors connected to screens and virtual realities rather than in the real world outside' (Schwab & Dustin, 2014, 27): 'Nature-deficit disorder is not a medical condition; it is a description of the human costs of alienation from nature. This alienation damages children and shapes adults, families, and communities' (Louv, 2005, 1).

Researchers have documented a variety of benefits tied to time in the outdoors, whether that be viewing pictures of the outdoors, being surrounded by nature or being engaged in activities intertwined with the natural environment (Yeh, Stone, Churchill, Wheat, Brymer & Davids, 2016). Benefits include improved mental health (Barton & Pretty, 2010), physical health (Andre, Williams, Schwartz & Bullard, 2017), including enhanced exercise adherence and motivation (Gladwell, Brown, Wood, Sandercock & Barton, 2013), academic success (Louv, 2005), emotional and social well-being (Greenleaf, Bryant & Pollock, 2014) and pro-environmental values (Louv, 2005). Researchers have also documented the benefits of physical activity in a natural versus a synthetic or built environment (Bowler, Buyung-Ali, Knight & Pullin, 2010; Shanahan, Franco, Lin, Gaston & Fuller, 2016). They have also found that time spent outdoors is associated with an increase in physical activity (McCurdy, Winterbottom, Mehta & Roberts, 2010).

The benefits of outdoor activities were more recently identified for Canadian youth aged 7–14. Focusing on ways to address the physical inactivity crisis among youth, researchers confirmed 'positive associations between time outdoors and physical activity' (Larouche, Garriguet, Gunnell, Goldfield & Tremblay, 2016, 3). They found that 'Children who report more time outdoors are more physically active and less sedentary, and display enhanced psycho-social health, compared with those who spend less time outdoors' (p. 13). Such findings have led to the encouragement of 'green exercise.' 'Green exercise, in its simplest form, is exercise performed in (relatively) natural environments such as parks,' according to Graham and Neill (2010, 239). Physical activities in the outdoors are thus becoming a prescribed remedy in North America to address societal issues such as sedentary behavior.

Adopting a Strengths and Hope perspective analysis on this issue, activities in the outdoors are currently being viewed in North America as a resource that can further existing strengths or create new ones, in which individuals engage in meaningful physical activities that promote 'healthy holistic child development.' Benefits ensuing from such strengths include combating depression, obesity and ADD; enhancing problem solving, critical thinking and decision making; stimulating creativity, and fostering stewardship of the environment (Louv, 2005). In a later book, Louv (2011) identified several benefits adults also gain by tapping into the restorative powers of nature. As discussed in this chapter, one way to address this challenge, or preferred vision, is by learning from/drawing on Indigenous peoples' understandings/strengths tied to land-based practices that can inform and enhance PAPs for all individuals. This vision aligns with elements included in *A Framework for Recreation in Canada 2015: Pathways to Wellbeing* (Canadian Parks and Recreation Association/Interprovincial Sport and Recreation Council, 2015, 22–25), as outlined below:

- Goal 2: Increase inclusion and access to recreation for populations that face constraints to participation, i.e., 'valuing cultural, ethnic and racial diversity is vital to the prevention of prejudice and discrimination.'
- Priority 2.4: 'Recognize and enable the experience of Aboriginal peoples in recreation with a holistic approach drawn from traditional values and culture.'
- Goal 3: Help people connect to nature through recreation.
- Priority 3.3: 'Develop public awareness and education initiatives to increase understanding of the importance of nature to wellbeing and child development, the role of recreation in helping people connect to nature and the importance of sustainability in parks and recreation.'

Conclusion

In this chapter, we have built on the argument that Indigenous traditional physical activities include land-based cultural practices, and that incorporating these practices within SDP programs and community physical activity programs can serve to legitimize a cooperative and land-focused logic that informs the physical activity practices of both Indigenous and non-Indigenous peoples. Such land-focused activities would encourage time outdoors and connections to the land, and directly address the potentially negative implications of the weakening bonds between mainstream Canadian youth and the natural environment. Further, these land-focused activities would avoid reinforcing the competition-based logic underpinning mainstream sport, and validate a less mediated environment for physical activity, as might be found in nature. In keeping with a Strengths and Hope perspective, we suggest that Dene and Inuvialuit traditional cultures that use land-based community practitioners in the Canadian Territories can serve as instructive examples, ones that youth can draw upon to strengthen ties to the land, and to potentially opt for a non-competition-based logic in their approach to physical activity.

The case described here also offers an instructive perspective on discussions concerning the environmental position of sports within the SDP sector. With some exceptions, the spatial constitution of dominant sports practices creates environmental differences in principle; thus the articulation of sport spaces to their embedding environments is often conceptualized in terms of remediation or the minimization of impacts. The ecological embeddedness of Indigenous PAPs, by contrast, would shift the concern to ecological relations as confirmation of cultural practices and identities. This practical reproduction of culturally specific connections to the land could in turn serve as affirmation of Indigenous positions of knowledge and authority, particularly in the context of SDP initiatives that often rely on external authority. For Indigenous peoples in Canada who seek to reclaim and confirm their cultural identities under conditions of marginalization and displacement, physical activity practices that place the land at their center can serve to uncover the inherent strengths of Indigenous culture. They simultaneously might offer an alternative model of cultural practice for SDP initiatives, potentially fostering an attachment to the land that promotes individuals' long-term commitment to addressing all forms of environmental degradation.

Notes

1 The term 'Aboriginal' used in the TRC's *Final Report* is equivalent to the term 'Indigenous' used in this chapter; we acknowledge that different politics inform these two usages.
2 Needless to say, this statement does not apply to all sports to the same degree, but it does hold true with regard to those organized sports that are popular in many Indigenous communities across Canada

References

Aitken, A. & Radford, L. (2018). Learning to teach for reconciliation in Canada: Potential, resistance and stumbling forward. *Teaching and Teacher Education*, 75, 40–48.

Andre, E.K., Williams, N., Schwartz, F. & Bullard, C. (2017). Benefits of campus outdoor recreation programs: A review of the literature. *Journal of Outdoor Recreation, Education, and Leadership*, 9(1), 15–25.

Bale, J. (1993). The spatial development of the modern stadium. *International Review for the Sociology of Sport*, 28(2–3), 121–133.

Bale, J. (1994). *Landscapes of modern sport*. Leicester, UK: Leicester University Press.

Barton, J. & Pretty, J. (2010). What is the best dose of nature and green exercise for improving mental health? A multi-study analysis. *Environmental Science & Technology*, 44(10), 3947–3955.

Bennett, J. & Rowley, S. (2004). *Uqalurait: An oral history of Nunavut*. Montreal, Canada: McGill-Queen's University Press.

Bowler, D.E., Buyung-Ali, L.M., Knight, T.M. & Pullin, A.S. (2010). A systematic review of evidence for the added benefits to health of exposure to natural environments. *BMC Public Health*, 10(1), 456.

Canadian Broadcasting Corporation. (June 2018). Beyond 94. Truth and reconciliation in Canada. Retrieved October 18, 2018 from https://newsinteractives.cbc.ca/longform-single/beyond-94

Canadian Parks and Recreation Association/Interprovincial Sport and Recreation Council. (2015). *A framework for recreation in Canada 2015: Pathways to wellbeing*. Ottawa, Canada: Canadian Recreation and Parks Association. Retrieved May 16, 2018 from https://static1.squarespace.com/static/57a2167acd0f68183878e305/t/5926efacebbd1a74b7b584d8/1495723950196/Framework+For+Recreation+In+Canada_2016+w+citation.pdf

Christian, J. & Gardner, P.M. (1977). *The individual in northern Dene thought and communication: A study in sharing and diversity*. Canadian Ethnology Service, Mercury Series 35. Ottawa, Canada: National Museum of Man.

CIRNAC. (2018a). Sport and reconciliation, updated October 25. Retrieved October 26, 2018 from www.aadnc-aandc.gc.ca/eng/1524505883755/1524505915831

CIRNAC. (2018b). Delivering on truth and reconciliation commission calls to action, updated October 26. Retrieved October 26, 2018 from www.aadnc-aandc.gc.ca/eng/1524494530110/1524494579700

Cruikshank, J. (1977). *My stories are my wealth*. Whitehorse, Canada: Council for Yukon Indians.

Cruikshank, J. (2005). *The social life of stories: Narrative and knowledge in the Yukon Territory*. Vancouver, Canada: UBC Press.

Dechinta Centre for Research and Learning. (n.d.). In community, on the land, for the future. Retrieved October 11, 2018 from http://dechinta.ca/what-dechinta-offers

Donnelly, P. (1996). Prolympism: Sport monoculture as crisis and opportunity. *Quest*, 48, 25–52.

Enright, M. (2014). CBC *The Sunday Edition* podcast with Tali Sharot and Denise Larsen, February 9. Retrieved July 10, 2019 from www.cbc.ca/radio/thesundayedi tion/cellphone-addiction-socks-for-the-homeless-colm-feore-on-lear-muqtida-mansoor-mail-pete-seeger-loving-the-beatles-mail-traffic-safety-the-science-of-hope-and-optimism-1.2904958

Gardam, K., Giles, A.R. & Hayhurst, L.M.C. (2017). Sport for development for Aboriginal youth in Canada: A scoping review. *Journal of Sport for Development*, 5(8), 30–40.

Gladwell, V.F., Brown, D.K., Wood, C., Sandercock, G.R. & Barton, J.L. (2013). The great outdoors: How a green exercise environment can benefit all. *Extreme Physiology & Medicine*, 2(1), 3.

Graham, J.M. & Neill, J.T. (2010). The effect of 'green exercise' on state anxiety and the role of exercise duration, intensity, and greenness: A quasi-experimental study. *Psychology of Sport and Exercise*, 11(3), 238–245.

Greenleaf, A.T., Bryant, R.M. & Pollock, J.B. (2014). Nature-based counseling: Integrating the healing benefits of nature into practice. *International Journal for the Advancement of Counselling*, 36(2), 162–174.

Guttmann, A. (2004). *From ritual to record: The nature of modern sports*. Updated edition. New York: Columbia University Press.

Hayhurst, L. & Giles, A. (2013). Private and moral authority, self-determination, and the domestic transfer objective: Foundations for understanding sport for development and peace in Aboriginal communities in Canada. *Sociology of Sport Journal*, 30(4), 504–519.

Health Canada. (2016). *A capture of national land-based initiatives funded by FNIBH programming*. Ottawa, Canada: Health Canada, Government of Canada.

Heine, M. (2012). *The culture and practice of Dene games*. Whitehorse, Canada: Transversant.

Heine, M. (2013). Performance indicators: Aboriginal games at the Arctic Winter Games. In J. Forsyth & A. Giles (eds.). *Aboriginal peoples and sport in Canada: Historical foundations and contemporary issues*. Vancouver, Canada: UBC Press, 160–181.

Heine, M. & Scott, H.A. (1994). Cognitive dichotomies: 'Sports,' 'games' and Dene cultural identity. *Communication & Cognition*, 27(3), 321–336.

Heine, M., Andre, A., Kritsch, I. & Cardinal, A. (2007). *Gwichya Gwich'in Googwandak: The history and stories of the Gwichya Gwich'in*. 2nd edition. Tsiigehtshik, Canada: Gwich'in Social and Cultural Institute.

Helm, J. (2000). *The people of Denendeh: Ethnohistory of the Indians of Canada's North-west Territories*. Montreal, Canada: McGill-Queen's University Press.

Honigman, J.J. (1968). Interpersonal relations in atomistic communities. *Human Relations*, 27, 220–229.

Honigman, J.J. (1975). Psychological traits in Northern Athapaskan culture. In A. McFadyen Clark (ed.). *Proceedings of the Northern Athapaskan Conference 1971*. Vol. 2. Ottawa, Canada: National Museum of Man, 546–576.

Jacobs, D. (2005). What's hope got to do with it? Toward a theory of hope and pedagogy. *Journal of Advanced Composition*, 25(4), 783–802.

Jacobs, D. (2008). The audacity of hospitality. *Journal of Advanced Composition*, 28(3/4), 563–581.

Kelly, L. (2018). 'Indoor generation': A quarter of Americans spend all day inside, survey finds. *The Washington Times*, May 15.

Larouche, R., Garriguet, D., Gunnell, K.E., Goldfield, G.S. & Tremblay, M.S. (2016). Outdoor time, physical activity, sedentary time, and health indicators at ages 7 to 14: 2012/2013 Canadian Health Measures Survey. *Statistics Canada Health Reports*, 27(9), 3–13.

Lavalée, L. & Lévesque, L. (2013). Two-eyed seeing: Physical activity, sport, and recreation promotion in Indigenous communities. In J. Forsyth & A.R. Giles (eds.). *Aboriginal peoples and sport in Canada*. Vancouver, Canada: UBC Press, 206–228.

Louv, R. (2005). *Last child in the woods: Saving our children from nature-deficit disorder*. Chapel Hill, NC: Algonquin Books.

Louv, R. (2008). *Last child in the woods: Saving our children from nature-deficit disorder*. Revised and updated edition. Chapel Hill, NC: Algonquin Books.

Louv, R. (2011). *The nature principle: Human restoration and the end of nature-deficit disorder*. Chapel Hill, NC: Algonquin Books.

McCurdy, L.E., Winterbottom, K.E., Mehta, S.S. & Roberts, J.R. (2010). Using nature and outdoor activity to improve children's health. *Current Problems in Pediatric and Adolescent Health Care*, 40(5), 102–117.

McGregor, H.E. (2012). Curriculum change in Nunavut: Towards Inuit Qaujimajatuqangit. *McGill Journal of Education*, 47(3), 285–302.

National Centre for Truth and Reconciliation. (2018a). Imagine a Canada. Retrieved October 18, 2018 from https://education.nctr.ca/imagineacanada/

National Centre for Truth and Reconciliation. (2018b). Memorial register. Retrieved October 18, 2018 from https://education.nctr.ca/memorial-register

Nicholls, S., Giles, A.R. & Sethna, C. (2010). Perpetuating the 'lack of evidence' discourse in sport for development: Privileged voices, unheard stories and subjugated knowledge. *International Review for the Sociology of Sport*, 46(3), 249–264.

Nunavut Department of Education. (2007). *Inuit Qaujimajatuqangit: Education framework for Nunavut curriculum*. Iqaluit, Canada: Department of Education, Government of Nunavut.

NWT Department of Education. (1993). *Dene Kede: The curriculum from the Dene perspective*. Yellownkife, Canada: Department of Education, NWT Government.

NWT Department of Education. (1996). *Inuuqatigiit. The curriculum from the Inuit perspective*. Yellowknife, Canada: Department of Education, NWT Government.

Paraschak, V. (2013). Hope and strength(s) through physical activity for Canada's Aboriginal peoples. In C. Hallinan & B. Judd (eds.). *Native games: Indigenous peoples and sports in the post-colonial world*. Bingley, UK: Emerald Publishing, 229–246.

Paraschak, V. & Thompson, K. (2014). Finding strength(s): Insights on Aboriginal physical cultural practices in Canada. *Sport in Society*, 17(8), 1046–1060.

Redvers, J. (2016). Land-based practice for Indigenous health and wellness in the Northwest Territories, Yukon and Nunavut: Plain language research summary. Yellowknife, Canada: Institute for Circumpolar Health Research, October. Retrieved July 7, 2019 from www.ichr.ca/wp-content/uploads/2016/11/Land-based -Research-Summary_2016.pdf

Saleebey, D. (2013). *The strengths perspective in social work practice*. Sixth edition. Hoboken, NJ: Pearson Education.

Schwab, K. & Dustin, D. (2014). Engaging youth in lifelong outdoor adventure activities through a nontraditional public school physical education program. *Journal of Physical Education, Recreation and Dance*, 85(8), 27–31.

Shanahan, D.F., Franco, L., Lin, B.B., Gaston, K.J. & Fuller, R.A. (2016). The benefits of natural environments for physical activity. *Sports Medicine*, 46(7), 989–995.

Smith, S.E.K. & Taunton, C. (2018). Pedagogies of remembrance and 'doing critical heritage' in the teaching of history: Countermemorializing Canada 150 with future teachers. *Journal of Canadian Studies*, 52(1), 217–248.

Styres, S.D. (2017). *Pathways for remembering and recognizing Indigenous thought in education: Philosophies of Iethi'nihsténha Ohwentsia'kékha (Land)*. Toronto, Canada: University of Toronto Press.

Truth and Reconciliation Commission of Canada. (2015). *Final report of the TRC*. Toronto, Canada: Lorimer.

United Nations. (2003). *General Assembly Resolution 58/5: Sport as a means to promote education, health, development and peace*. November 3. Retrieved on May 16, 2018 from https://documents-dds-ny.un.org/doc/UNDOC/GEN/N03/453/21/PDF/N0345321.pdf

Wamsley, K. (2002). The global sports monopoly. *International Journal*, 57(3), 395–410.

Wildcat, M., McDonald, M., Irlbacher-Fox, S. & Coulthard, G. (2014). Learning from the land: Indigenous land based pedagogy and decolonization. *Decolonization: Indigeneity, Education & Society*, 3(3), 1–14.

Yeh, H.P., Stone, J.A., Churchill, S.M., Wheat, J.S., Brymer, E. & Davids, K. (2016). Physical, psychological and emotional benefits of green physical activity: An ecological dynamics perspective. *Sports Medicine*, 46(7), 947–953.

Index